12/03

The Art of Fine Enameling

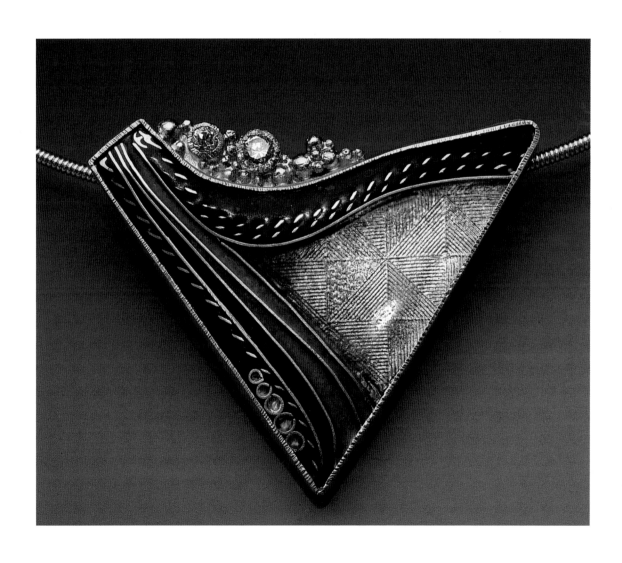

Homage Basse Taille on copper

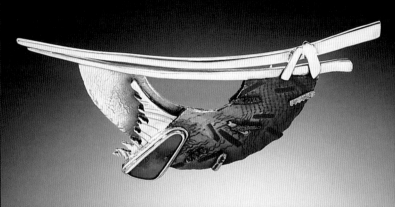

Towards East Champlevé on stainless steel with 22k, 18k, 14k gold

Cityscape Sculpture

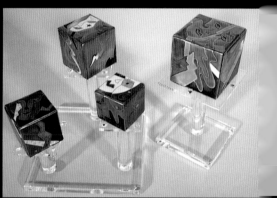

Mask-erade

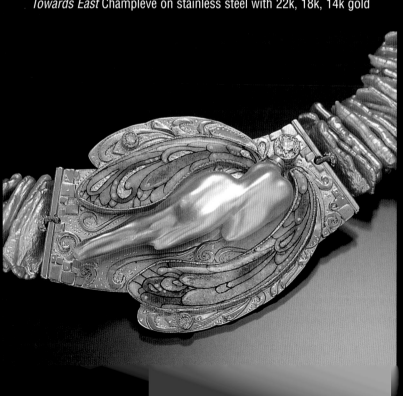

The Art of Fine Enameling

Karen L. Cohen

Sterling Publishing Co., Inc. New York

A Sterling/Chapelle Book

Chapelle, Ltd.:

- Owner: Jo Packham

- Editor: Ray Cornia

- Technical Editors: Tom Ellis, j.e.jansen

- Staff: Areta Bingham, Kass Burchett, Jill Dahlberg, Marilyn Goff, Karla Haberstich,
  Holly Hollingsworth, Susan Jorgensen, Barbara Milburn, Karmen Quinney,
  Caroll Shreeve, Cindy Stoeckl, Kim Taylor, Sara Toliver, Desirée Wybrow

If you have any questions or comments, please contact:
Karen L. Cohen at karen@kcEnamels.com or http://www.kcEnamels.com
Questions can also be referred to:
Chapelle, Ltd., Inc., P.O. Box 9252, Ogden, UT 84409
(801) 621-2777 • (801) 621-2788 Fax • e-mail: chapelle@chapelleltd.com
website: www.chapelleltd.com

Photo on half title page by Bob Barrett. Cloisonné with guilloché necklace, 18k & 24k, gold by Marilyn Druin

Library of Congress Cataloging-in-Publication Data

Cohen, Karen L.
  The art of fine enameling/Karen L. Cohen
    p. cm.
  Includes index.
  ISBN 0-8069-7869-4
    1. Enamel and enameling—Technique. I. Title

NK5000.C64 2002
738.4dc21   2002066867

10 9 8 7 6 5 4 3 2 1

Published by Sterling Publishing Co., Inc.
387 Park Avenue South, New York, NY 10016
©2002 by Karen L. Cohen
Distributed in Canada by Sterling Publishing
c/o Canadian Manda Group, One Atlantic Avenue, Suite 105
Toronto, Ontario, Canada M6K 3E7
Distributed in Great Britain and Europe by Chrysalis Books
64 Brewery Road, London N7 9NT, England
Distributed in Australia by Capricorn Link (Australia) Pty. Ltd.
P.O. Box 704, Windsor, NSW 2756, Australia
Printed in China
All Rights Reserved

Sterling ISBN 0-8069-7869-4

Earth Beneath Our Feet incense burner by Harlan W. Butt

Foreword

Enameling, the technique of fusing colored glass powder onto a metal surface under high heat, has long been associated with jewelry and ecclesiastical objects. In the Middle Ages, glowing enamels were often used in place of gems. Court jewelers from the renaissance through the 18th century embellished elaborate brooches, watches, and clothing ornaments with enameled portraits and delicate floral bouquets, and enhanced set gemstones with enameled gold surfaces. The elaborate Easter eggs, cigarette cases, and personal accessories created by Carl Peter Fabergé in the 19th century brought brief popularity again to enameling, as did the fascination with color and surface in high-style French jewelry during the 1920s and '30s.

Since that time, however, enameling has been thought of popularly as a hobby craft predominated by ashtrays and lightswitch plates decorated by amateurs enameling precut copper shapes in home kilns. Some fifty years ago, Kenneth Bates, an author-craftsman in Cleveland, Ohio, publicized the creative possibilities of enamel to other craftsmen through his book on the process. Since that time and that book, technical advances have been matched by creative application of them.

Today, glass fired onto metal is no longer limited to jewelry, decorative souvenir spoons, or luxurious personal accessories. Large kilns to fire enamel onto sinks and bathtubs—or even exterior walls of buildings (think of midcentury filling stations and Richard Meyer's contemporary architecture) are now also the tools of artists. "Paintings" enameled onto metal panels grace building exteriors and public spaces, impervious to weather and resistant to vandalism.

Enameling can be undertaken by almost anyone, and Karen L. Cohen shows how! With straightforward descriptions of the tools and materials, and how to use them, she encourages even novices to enamel a variety of time-honored techniques. The step-by-step processes to make a variety of enameled objects will introduce both established techniques, and some newer uses of them, to the student. The easy-to-understand explanations will undoubtedly acquaint even experienced enamelists with the artistic and technical possibilities of enamel.

The biographies of the artists who share their techniques provide personal introductions to not only a Who's Who of enamelists practicing today. They, along with the Gallery section of the book, also reveal the myriad possibilities for individual expression once the process is understood and experienced. I commend the author and hope that devotees of this time-honored medium will—as I did—find it fascinating, too.

Lloyd E. Herman

Lloyd E. Herman was the founding Director of the national craft museum of the United States, the Smithsonian Institution Renwick Gallery in Washington, D.C., from 1971 until 1986. He also planned and directed the Canadian Craft Museum in Vancouver, Canada, 1988–90. He has written about craft arts in all media, and organized traveling exhibitions to bring attention to such vital aspects of American visual arts as enameling.

Preface

Welcome to the wonderful world of enamels! A world so diverse, practically no other medium can compare. Not only is there a variety of techniques to use, but also a wide range of what can be enameled.

But what is enameling? In its easiest terms, enameling is the fusing of glass to metal under high-temperature conditions. Thus, anything that can be made from metal can be enameled; anything from vases to spoons, from frames to jewelry, from metal mesh to solid forms, from large to small, from fancy to plain. Even the tub inside your common household washing machine is most likely enameled.

This book is a look into the beauty of fine enameling, where artwork is created with colors and textures that bring joy to the heart. To start, why not simply browse the tantalizing photos that are included. You will get a feel for the diversity of the medium and the charm of the finished pieces. And unlike painting and fabric arts, enamels won't fade with time; they are as permanent a medium as you will find.

The projects included in this book give a survey of techniques, from the traditional, such as cloisonné, champlevé, and plique-à-jour, to the more experimental, like raku. Each project was created and written by the individual artist, using his or her personal techniques. These are meant to be a starting point for your own creations.

Enameling has been an art form for centuries. Much has been written and shown of the old masters, like Fabergé and Lalique, and from the "Limoges" school. But what about today's enamelists? Although not great in number, they are great in workmanship. Thus, I have made *The Art of Fine Enameling* a celebration of modern enamelists whose work is truly wonderful.

So, get ready to learn a few things and to enjoy the view!

Karen L. Cohen

John Killmaster

Cry, Oh Eye of the Rain forest for we will not see Enamel on repoussé steel

Homage Basse Taille Torch-fired doors

Harold Balazs

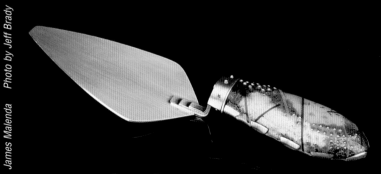

James Malenda   Photo by Jeff Brady

Dessert Trowel Limoges with decals

Table of Contents

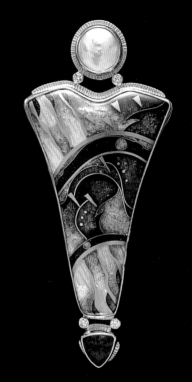

Ricky Frank   Photo by Ralph Gabriner

Barbara Minor   Photo by Ralph Gabriner

Enamel beads

Alana Clearlake   Photo by Kate Cameron

Fall: Fallen Leaf Plique-à-jour copper mesh

Enameling, Then and Now

Photo #1

Photo #2

Photo #3

Karin Pohl   Photo by G. Post

Mona & Alex Szabados   Photo by G. Post

Fredricka Kulicke   Photo by Ralph Gabriner

The earliest known enameled pieces have been dated to the13th century BC, when Mycenaean goldsmiths inlaid enamels into gold rings. Since then, cultures all over the world have incorporated enameling into their art forms. In the 5th century BC, Greek artisans used enamel to decorate artwork such as the Phidias statues of Zeus.

In the 9th through 11th centuries AD, gold cloisonné was popularized in the Byzantine Empire. In Germany, enameling was introduced by a Byzantine princess, Theophano, who married the German King Otto II, and then brought craftsmen with her from her native land. Thus German enamels were heavily influenced by these Byzantine roots. In Western Europe in the 12th century, champlevé religious objects were produced. And Renaissance goldsmiths, like Cellini, created beautiful enameled pieces in basse taille and plique-à-jour.

A revolution in enameling occurred in the late 15th century when the family of Pénicaud innovated a new method of "painting" with enamels. As this was developed in the French town of Limoges, the method is called limoges. This was the first time that enamel colors touched each other without the use of separating wires or metal. Using this method, portraits and other scenes could realistically be reproduced. The art of portraiture was highly developed by the great 16th century enamelists Leonard Limousin and Pierre Reymond.

Subjects and objects used for enamels are wide spread, but religious themes and objects have always been a favorite. See Photo #1. Other common themes are flowers and animals. See Photo #2. Common objects are boxes, candlesticks, jewelry, and watches. Enameled watches were first introduced at the end of the 16th century. Enamels have also been used instead of gemstones in precious jewelry and other objects. For example, in medieval times, a style of jewelry, called the garnet jewel, was popular. In these pieces, garnet was cut flat and inlaid between a metal framework that had been soldered to a metal backing, which was then repousséd. Today, this construction is done using enamels instead of actual gemstones. See Photo #3.

The Arts and Crafts movement, in modern times, has made a large impact in enameling. Kenneth F. Bates, an American university educator, moved enameling out of "hobby art" and is credited with influencing a multitude of modern enamelists. Schools like the Kulicke-Stark Academy of Jewelry Arts in New York City in the 1970s trained many enamelists and instilled a love of the medium. Fred Ball, with his innovative approaches to enameling, helped many students to stretch their imagination in working with enamels. Today there are craft schools and universities around the world teaching enameling. The most well-known organization of enamelists in the United States is The Enamelist Society, which sponsors conferences every other year, publishes a magazine, and has members in countries around the world. Local guilds exist, but the number of enamelists still remains small.

Through the years, a variety of enameling techniques has been developed. Some involve how the metal is prepared and some involve how the enamel is applied. The following defines the most prevalent, but by no means all, techniques:

• **Basse Taille:** French for "low cut." A technique in which a pattern is created in the metal backing before enameling. See Metal Patterning for Basse Taille on page 26 for information on how to get patterns onto the metal. See Photo #4. Also see the Basse Taille project on page 42.

• **Camaieu:** also called "en camaieu," a term dating from the mid-18th century describing a grisaille-like technique which uses a buildup of white enamel to create highlights and light areas. However, instead of using a black background, as in grisaille, transparent enamel is laid in first, beneath the whites. This technique is frequently used on snuffboxes, watches, and medallions. See Photo #5.

• **Champlevé:** French for "raised field" or "raised plain." A technique in which enamel is inlaid into depressions in the metal, leaving metal exposed. The depressions are typically made by an etching process, although other methods exist. First done in the 3rd century AD by the Celts decorating their shields, this technique has been one of the favorite forms of enameling. See Photo #6. Also see the Champlevé Panel project on page 48.

• **Cloisonné:** French for "cloison" or "cell." A technique in which metal wires are bent to form a design; enamel is then inlaid into the resulting "cloisons." Although this can be done in copper, contemporary cloisonné is most frequently done in silver or gold. The Byzantine Empire, 6th century AD, was the setting for gold cloisonné pieces of a religious nature. In the same time frame, the Japanese were producing scenes of nature. In China, cloisonné has been used since the 13th century AD. See Photo #7. Also see the Cloisonné Brooch project on page 52.

• **Ginbari Foil:** a technique, developed in Japan, using a foil design made with an embossing plate. This is an excellent technique for reproducing a design, as the embossing plate is reusable. It somewhat has the look of cloisonné; however, the "lines" are not wire, they are embossed foil. See the Ginbari Foil Embossing project on page 68.

• **Grisaille:** French for "greyness." A form of "painting" with enamel in a monochrome, using a black background with a buildup of white overlays. See the Grisaille project on page 74.

• **Guilloche:** French for "engine-turning." Engine-turning is the mechanical cutting of lines on metal to create a design. Because the pattern is engraved, the reflection of light through the overcoating of transparent enamel is enhanced, and its brilliance can be seen as the piece is moved from side to side.

Ray Parisi   Photo by Ralph Gabriner

Photo #4

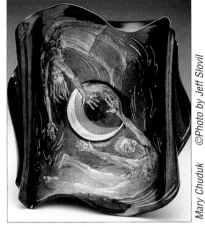

Mary Chuduk   ©Photo by Jeff Slovil

Photo #5

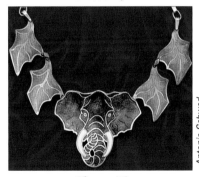

Antonia Schwed

Photo #6

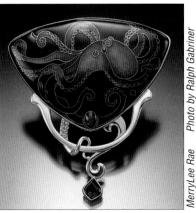

MerryLee Rae   Photo by Ralph Gabriner

Photo #7

Photo #8

Photo #9

Photo #10

The best known, but not the first, artist using this technique was Fabergé, in Russia, who, when showing pieces in Paris in 1900, brought a new interest to this technique. Guilloche was a dying art until Pledge & Aldworth Engine Turners, an English firm, revived it in the 1970s. See Photo #8. This is a sample of guilloche and cloisonné combination.

- **Impasto:** a technique in which acid-resistant painting enamel is applied to a bare metal surface, then fired. Multiple layers can be worked to build up a relief design, which can be sculptural in effect. Finally, the piece is covered with a transparent color. Other colors then can be added in thin layers only.

- **Limoges:** a technique of "painting" with enamel in which different enamel colors are put next to each other without the separation of wire or surface metal. See Photo #9. Also see the Limoges–Painting with Enamels project on page 78.

- **Plique-à-jour:** French for "membrane through which passes the light of day." A technique that resembles miniature stained glass and is reminiscent of its Art Nouveau and old-world influences. There are two basic methods of plique-à-jour: surface-tension-enameled and etched-enameled.

Photo #11

Photo #12

The surface-tension-enameled method has two different styles of metal construction: the first is pierced. See Photo #10. Also see the Plique-à-jour Pierced-heart Pendant project on page 102. The second style is filigree or skeletal framework. See Photo #11. The filigree style was first used in the 11th century and accepted all over the world.

The etched-enameled method is called Shōtai-Jippō, and sometimes "crystallized cloisonné" in Japan. See Photo #12. It is done somewhat like cloisonné with a copper backing and silver wires, but after the piece has been finished, the copper backing is etched off. Plique-à-jour pieces, because of the open back, are more fragile than other types of enamels.

• **Raku:** a technique in which hot enamel that includes oxides is put through a reduction firing, resulting in iridescent colors. The technique can be used with or without silver nitrate crystals. See Photo #13 for an example without silver nitrate. See the Raku-fired Bowl project on page 108 for raku done with silver nitrate.

• **Sgraffito:** a technique in which lines are drawn through a layer of unfired enamel, exposing the fused enamel (or bare metal) underneath. See Photo #14, an example of sgraffito and silkscreening. See the Sgraffito Plate project on page 118.

• **Silkscreen:** a technique in which designs on material mesh, such as silk, polyester, or nylon, are transferred onto an enameled base; this is similar to silkscreening on cloth. See Photo #14. Also see the Silkscreen for Enameling project on page 126.

Photo #13

Leni Fuhrman

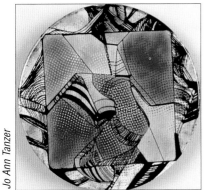
Photo #14

Jo Ann Tanzer

Photo #15

Harold Balazs

• **Stenciling:** a technique in which a design is cut into a material, such as paper or Mylar®, through which the enamel is applied to, or removed from, the metal. Thus, the "holes" that are cut can be either the positive or the negative space of the design. That is, one can sift enamel onto the metal, lay down the stencil, then use a brush to remove the enamel in the cut-out area (negative). See Photo #15. Or, the stencil is laid on the metal and enamel sifted into the cut-out area (positive). See the Stenciled Tile project on page 121.

• **Torch-fired:** a method of enameling in which a torch is used for the heat source, instead of a kiln. See Photo #16. Also see the Torch-fired Beads project on page 142.

Although each of these techniques can be used by themselves, two or more can be combined in one piece.

In addition, enameled pieces can be enhanced by decorative additives such as:

• **China Paints:** low-fire compatible ceramic materials that can be used on the top surface of enamels. See the Decals project on page 58.

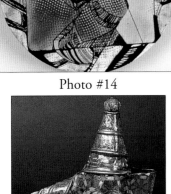
Photo #16

Deborah Lozier

• **Overglazes and Underglazes:** finely ground pigments used either over or under the regular enameling layers. Underglazes are particularly effective in a basse taille design. See the Limoges–Painting with Enamels project on page 78.

• **Copper Screen or Pot-scrubber Mesh:** elements for use on top or under transparent enamels, giving a wonderful texture to a piece. The screen can be used to give an interesting grid effect. See Photo #17. If used slightly under the enamel surface, when the surface is ground down, screening can give the effect of woven fabric as the stoning picks up the high parts where the warp wire crosses over the weft wire, leaving copper glints that give a textured pattern on the surface.

Photo #17

Jean Tudor

j.e. jasen

Photo #18

D. Rooke-Harris

Photo #19

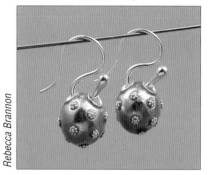

Rebecca Brannon

Photo #20

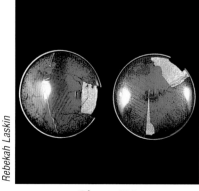

Rebekah Laskin

Photo #21

• **Decals:** designs or pictures printed on specially prepared paper for transferring an image to enamel, glass, wood, etc. See Photo #18. See the Decals project on page 58.

• **Foil and Leaf:** come in both fine silver and gold. In addition, leaf, which is much thinner, also comes in palladium. These elements can be placed under the enamel or on a top layer. See Working with Foils on page 36. Special foil objects, called paillons, are small preshaped designs that have a slight relief to them. See Photo #19. Various projects in this book use foil.

• **Gemstones:** can be added in an enameled area, using a metal bezel, which adds relief to a flat piece. See the Stone Setting Within an Enamel project on page 130.

• **Granules:** small grains of fine silver or 22k or 24k gold that can be used for top-layer embellishment of an enameled piece. See the Cloisonné Brooch project on page 52. Note that these small balls can be "granulated" (fused) to a thin back plate and then enameled around. See Photo #20.

• **Glass Beads and Balls:** can be purchased without holes and fused to the top of enameled pieces. See the Liquid Enamel and Glass Ball Additives project on page 88.

• **Lumps and Threads:** "lumps" are odd-formed chunks of colored glass and "threads" are filaments (short or long, thin or thick) of colored glass. Each can be fused into an enameled piece. See Torch-fired Beads project on page 142.

• **Lusters:** metal colors thinly applied on the top layer of an enameled piece. These sometimes fire with a crackle-maze effect, allowing the enamel underneath to show through. Some fire iridescent and some opalescent.

• **Metal:** small pieces of shaped metal can be added on the top layer of an enameled piece. They are embedded in a similar way to granules.

• **Millefiori:** cross sections of glass canes that include intricate patterns. Millefiori is best known in Venetian glass objects such as vases, paperweights, and lamps.

There are some methods of enameling that do not fit into either a technique or a decorative additive, but are a combination of the two. These include:

• **Firescale Enameling:** the use of the oxide buildup on a metal. Some pieces are completely done through firescale manipulation (by painting with a holding agent, sifting transparent enamels, and building up the resulting firescale lines), and some pieces are enhanced by the additive use of firescale (could be from a flaked-off piece that is reattached). See Photo #21. Also see the Sgraffito Plate project on page 118.

• **Separation Enameling:** a special type of enamel that when applied over regular enamel, indents the enamel and changes its color. See the Separation Enameling project on page 114.

In reality, any of these techniques, decorative additives, and methods can be combined to make a piece that is truly unique. Simply let the imagination soar.

Although there are various tools and supplies needed for the different enameling techniques, there is a set of common items that can be called an Enamelist's Tool Kit. The tools and supplies for different techniques are described in the appropriate projects. This section defines the general "tool kit," which is comprised of:

Safety and Protection

- Dust/particle mask
- Eye protection
- Studio layout
- Venting system
- Work clothes

Cleaning Metal

- Cleaning abrasives
- Glass brush or scrubbing pad
- Metal cleaners
- Pickling tools

Enameling Application

- Application tools
- Enamels, overglazes, and underglazes
- Holding agent

Firing

- Firing supports
- Heat source
- Iron planche (optional)
- Kiln wash
- Tools for handling hot items

Finishing

- Finishing tools

Miscellaneous

- Fire extinguisher
- Magnifying lenses (optional)
- Paper towels, tissues, and cotton swabs

Safety and Protection

Dust/particle mask

Enamels are fine particles and, whether leaded or unleaded, they are bad for the lungs. If using leaded enamels, be certain to purchase a mask that is OSHA rated for lead. See Photo #1. Also, particles can remain in the air for a while after sifting, so keep the mask on for an appropriate amount of time.

Eye protection

Infrared rays are emitted from a heated furnace. If one often looks directly inside, cataracts, a hazard of many glass blowers, could develop. Be certain to wear eye protection (safety glasses with welder's shade #2 lenses) and avoid looking too long. Eye protection is especially important when working with materials that may fly up and hit the eyes or face. A clear face shield is a protection from a variety of hazards.

Studio layout

It is best to have a separate enameling area from living quarters so that dirt and dust can be contained. It is not a good idea to use the kitchen or bedroom for enameling.

Venting system

Lusters and other applicants, oil-based painting enamels, and soldering emit fumes; sifting raises dust. These substances should not be in the air space as they are health hazards. Various types of venting systems are available. An easily obtainable one is an ordinary range hood, with an external exhaust vented through a furnace filter. Be certain that it vents an adequate CFM (cubic feet per minute) for the air space of the studio and that the hood captures the maximum amount of dust and fumes. See Photo #2.

Photo #1

Photo #2

Stephen Lanza

If sifting is frequently done, it may be best to use a venting system that pulls fine particles from below the sifting area. It can be constructed with a perforated plate or screen placed below the work area and connected to a suction system.

Work clothes

Use an apron and cap to cover and protect clothes and hair. Also, keep a set of clothes that is used only in the enameling studio. Different aprons may be needed for various activities, such as a heat-resistant apron for the kiln station and a plastic apron for water-related activities. Wear cotton as synthetics can melt.

Cleaning Metal

An enamelist must remove oil and grease from metal and remove fines from enamels. The basics include:

Cleaning Abrasives

Polishing stones or sandpaper are needed to either clean up the metal edges of the piece after firing or to finish the enamel. Alundum stones work well. See Finishing on page 37.

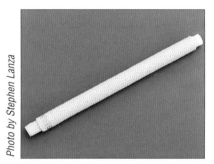

Photo #3

Glass brush or scrubbing pad

Usually, the enamel piece needs to be cleaned between firings and after stoning. This is easiest done with a glass brush or scrubbing pad. A glass brush is also used to prepare fine silver for enameling. See Photo #3. Because this tool is made with fiberglass, be careful not to get fibers in your fingers. Be certain to always use it under running water.

Metal cleaners

In addition to pickling, there are alternative ways to prepare metal for enameling. See Preparing Metal for Enameling on page 27.

Pickling tools

Photo #4

Metalsmithing nonferrous pickle solution is a mild acid used to remove firescale and, in general, to remove oxide from metal. It is used primarily after soldering and, when enameling on copper or sterling silver, between enamel firings. Commercial pickle comes as a powder that is mixed in water (**Note:** *Always add acid to water, not the reverse.*) Normally, warm pickle solution is used for plain metal, and cold pickle solution for enameled metal; special pickle pots are sold to hold and warm the solution. See Photo #4. It is important to keep a clean water bath near the pickle solution as all pieces must be rinsed after being pickled. Also, a pair of copper tongs is necessary to remove the metal from the pickle solution; noncopper metal tongs will contaminate the solution.

Notes: *It is best to avoid mixing metals in the pickle solution.*
As with all acids, treat it with respect and wear protective clothing.

Enameling Application

Application tools

These tools are used to apply the enamel onto the metal. Obviously, the tool chosen depends on the method of application—sifting and wet-packing are the most common. See photo #5.

For sifting, a sifter and something to hold the piece while sifting are necessary. A paintbrush to sweep away stray enamel after sifting. See Sifting on page 31.

For wet-packing, paintbrushes or a spatula can be used. See Wet-packing on page 32 and the Plique-à-jour Pierced-heart Pendant project on page 102.

Photo #5

Enamels, overglazes, and underglazes

This book deals with vitreous enamel. Vitreous is defined as an adjective meaning "of, relating to, derived from, or consisting of glass." Vitreous enamel can be defined as glass fused onto metal by means of high heat. Other materials, also called enamel (for example enamel house paint), are not of the vitreous type.

There is some confusion regarding the difference between jewelry and porcelain enamel. Both are vitreous and differ only in how they are manufactured and applied. The color in jewelry enamel is produced when the glass is smelted; the color in porcelain enamel is produced by milling a variety of materials, including various glasses and ceramic pigment, in a ball mill. These two enamels can be used in the same piece. Porcelain enamels are typically found in liquid form as clay is added as a base. The term "porcelain enamel" was first used by the industry to take advantage of the popularity of porcelain dinnerware. Thus, this term is only a sales pitch and should not denote a difference from other vitreous enamels.

Photo #6

Enamels come in various forms. See Photos #6 and #7. Enamels come in powder (in various mesh sizes), lump, liquid, crayon, pen, watercolor, painting, and acrylic. There are four types of enamels: transparent, opaque, translucent, and opalescent. As the names imply, transparent enamels allow what is underneath to show through; opaque enamels conceal whatever is underneath them; translucent enamels conceal, yet allow the passage of light; opalescent enamels are semiopaque with an opal sheen. All four come in an array of colors. Clear enamel is traditionally called flux, but more accurately called "transparent clear;" flux is a confusing term as it conflicts with soldering flux. The term "painting enamel" refers to finely ground enamel smaller than 325 mesh.

Photo #7

Enamels can also be either leaded or unleaded. It is important to note that for safety reasons, leaded enamels are no longer manufactured in the United States; however, enamel manufacturers in other countries still do.

Enamelists tend to use enamels from a single enamel company because enamels from different manufacturers can have different expansion and contraction characteristics that may not be compatible with each other. If using different brands of enamel, test-fire the desired combinations to determine compatibility.

Holding agents

A holding agent is a material used to help hold enamel and wires in place, both during the enamel application and during firing. It is basically an enamel "adhesive" and should be used sparingly. Holding agents dissipate in heat and leave no ash. Sometimes an agent is used full strength and sometimes diluted, and can be applied with a brush or a sprayer. The most commonly used holding agent is Klyr-fire®, which is sold by virtually all enamel suppliers. Other holding agents are enameling oil (used many times with gold leaf and foils), basque or lotus root (for enameling on a curved surface), gum arabic, and gum tragacanth. See Photo #8.

Firing

Firing supports

When placing a piece in the kiln, the enameled area needs to be protected from sticking to the wrong surface. Suspend the piece by its edges so that they are stable, unless by doing so the weight of the piece will cause it to sag. If prone to sag, support the piece from underneath with as little surface area of the support touching the piece as possible.

Many types of firing supports are available. Use what is best for the technique and piece being fired. Typical supports are trivets, wire-mesh screens, and mica. See Photo #9. Trivets come in various sizes and shapes; there are 3-point, 4-point, and sawtooth, to name a few. Wire-mesh screens can be used to hold a piece or to hold the trivet and make it easier to put the piece inside the kiln. Mica is a thin sheet of nonflammable material on which a piece is rested. Mica will stick to a piece (usually the counter enamel) and must be removed at the end of the firing.

Trivets and wire mesh are metal and must be cleaned of fused enamel bits and possible firescale, which can contaminate a piece. Enamel bits can be removed with an abrasive stone. Firescale can be removed by tossing it into cold water while it is hot. Alternately, scrape it with an abrasive stone or file.

Heat source

Because enamel is glass fused at high temperatures, a heat source is necessary. The two main sources are a furnace/kiln and a torch. Most of the projects in this book are fired using the furnace method. See the Torch-fired Beads project on page 142 for instruction on torch-firing.

Enamelists use what is called a "furnace." Furnaces and kilns are the same equipment, but differ in how they are used. A furnace is heated to temperature; the piece inserted and fired, then removed while still at a firing temperature. In a kiln, the piece is inserted when the kiln is at room temperature, the kiln is then heated to firing temperature, and the piece remains in the kiln until both have cooled. The terms furnace and kiln may be used interchangeably in this book.

Furnaces/kilns come in various sizes. See Photo #10. These are small tabletop models. See Photo #11. This is a medium size model. The size of the kiln's inside chamber limits the size of the piece that can be made. Jewelry can usually be done in the smallest kiln. And sometimes larger pieces can be constructed by creating

Photo #8

Photo #9

Photo #10

Photo #11

smaller pieces that fit together. See the Limoges–Painting with Enamels project on page 78 for an example. Furnaces can be much larger than the ones shown.

Some kilns have a removable tile that fits the bottom of the kiln. This is called a floor shelf and is used to ensure that the actual bottom of the kiln does not get messy. To make the floor shelf easy to clean, cover it with kiln wash. When the floor shelf gets dirty, scrape off the kiln wash, then reapply a new coat.

Iron planche (optional)

Sometimes the enamel piece will need to be flattened. This is done with a planche — a heavy piece of flat iron with a handle. See Enameling Troubleshooting section page 152.

Kiln wash

Kiln wash is used to coat items so that enamel does not stick to them. It comes in a powder form. Simply mix a small amount with some water until it becomes the consistency of cream. Use a paintbrush to apply, or pour on an even coat. Dry completely.

Tools for handling hot items

Since high heat is being used, proper tools are necessary for handling hot items. See Photo #12. A firing fork is needed to get items into and out of the kiln, heat-resistant gloves, a heat-resistant tile for the work area, plus other tools as necessary for the size of piece that is being worked on. One suggestion is to use stainless-steel test tube tongs from a local chemical supply house to hold hot items and to get items in and out of acid (but not pickle solution). See Photo #13.

Finishing

A selection of sandpaper, files, abrasive stones, power buffers, and polishing compounds are necessary to grind, smooth, and polish enamel pieces. See Finishing on page 37.

Miscellaneous

Magnifying lenses (optional)

Many times enamel will have to be applied into small areas or small imperfections in the work will need to be examined. Magnifying lenses or a jeweler's loop may help. Magnifying lenses come in various forms such as a clip-on for eye glasses, glasses themselves, a visor, or on a movable arm. Each of these optical aids accommodates personal choice and can be purchased in a variety of magnifying strengths. See Photo #14.

Paper towels, tissues, and cotton swabs

These items work well to wick water from wet enamels. Paper towels are useful to clean up spills and can be used as lightweight stencils.

Fire extinguisher

A regular household fire extinguisher is adequate; however, be certain it is large enough to be effective.

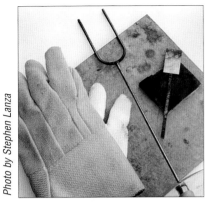

Photo #12

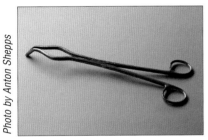

Photo #13

Photo #14

Studio Basics, Tips, and Tricks

Enamelists do not always agree on how things should be done; many create enamel pieces using different procedures for the same technique. Thus the information in this chapter should be viewed as merely a starting point.

Some things all enamelists do agree on include:

• Keep the work environment clean so that enamels do not get contaminated.

• Properly prepare enamels and metal before enameling begins.

These precautions will eliminate many problems before they occur.

The following section provides general studio information, tips, and tricks. It is compiled from scholarly texts, studio experience, and the experience of the project artists who have kindly contributed to this book. Additionally, see the Appendix on page 156 for further information.

Tip #1: A good place to get tools is to ask a dentist for tools scheduled to be thrown out, especially diamond drill bits (be certain to sterilize all dental tools).

Many different types of metal can be used for enameling; however, this section is geared towards copper and fine silver, the metals used in this book's projects. A fairly new material called metal clay becomes metal when fired. The silver form of this material can be enameled, just like any other fine silver, as long as the surface has been burnished. See Photo #1.

Mary S. Reynolds   Photo by Richard Brunck

Photo #1

Safety First

An enameling studio has fumes, dust, heat, chemicals, and other hazards. This section suggests some safety basics for avoiding health problems when enameling:

• See Safety and Protection on page 13.

• Don't eat or drink in the enamel studio—dangerous chemicals may be ingested.

• Don't allow pets in the studio; they can carry metal or enamel dust to other parts of the house.

• Don't allow small children in the studio unattended.

• If leaded enamels are used, check with a doctor on what personal lead levels should be. Take a baseline lead test, and then another once a year to ensure lead levels are not increasing.

Note: An acceptable lead level in an adult may be too high for a child. Also remember that lead poisoning in children can cause brain damage. If leaded enamels are used in a household with children, be especially careful to have good ventilation.

• When finished in the studio, be certain to wash well—enamel dust and fumes get on clothes, hair, face, arms, etc. Always wash yourself thoroughly and wash enameling work clothes separately from the rest of the laundry.

• Heat and hot objects are always a potential hazard. Be certain to have heatproof surfaces on which to place hot items, wear heat-resistant gloves when appropriate, avoid looking too much into the heat of the kiln, have a fire extinguisher handy, and keep Aloe Vera in the studio.

Tip #2: There is nothing better than Aloe Vera for burns. Aloe Vera gel (98% aloe) comes in a tube or pump bottle. When burned, immediately cover the burnt area generously with Aloe Vera gel. Cover some of the unburned skin around the burnt area as well. In approximately five minutes, the pain should subside. Keep the gel on for several hours or overnight. In the morning there may not be a blister or any pain. The aloe may be covered with a bandage; however, keep it loose. See Photo #2.

Firing

The firing temperatures for most enamels are between 1300°F and 1600°F (for most metals, not including aluminum). Common firing temperatures are from 1450°F to 1500°F. Each enamel color is different, so a test-firing should be done to determine what temperature is best. A kiln may have a pyrometer, but not all are calibrated correctly. Thus, it is important to be able to read the temperature by the color of the heat as follows:

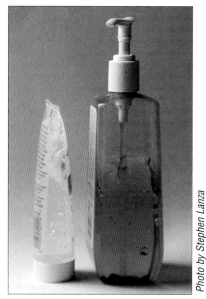

Photo #2

Color Description	Temperature
Dark Red	1300°F
Cherry Red	1400°F–1500°F
Light Orange/Red	1550°F–1600°F
Orange Yellow	Over 1600°F

Tip #3: To get the best color of a transparent or flux enamel directly on copper, the first layer should be fired high—approximately 1500°F–1600°F so that the oxides in the copper can be taken into the solution of the enamel. If necessary, fire the copper once, cool, then rotate the piece and fire again. This may be necessary because most kilns fire hotter in the closed-in back than the front where the door allows heat to escape.

Firing times for small pieces can be in the "one minute to several minutes" time frame. The larger the piece, the longer the firing time. As long as the kiln temperature does not go above the melting temperature of the metal, there is no need to worry about the piece melting. One exception to this is when a silver cloisonné wire is placed on a copper base; this combination has a lower melting point than either of the two individual metals. Enamel can burn if it is fired too high, or if it is fired for too long, or if the application of enamel is too thin. Firing takes practice and, with experience, the enamelist will learn how to control the heat.

Other factors which affect firing time are size of kiln, thickness of kiln insulation, accuracy of pyrometer, how much heat is lost from opening the door, size and thickness of piece, type of metal, size and thickness of trivet or stilt, color of the enamel, enamel particle size. See Photo #3. The firing time for the same-sized piece of steel will take longer than a piece of copper, which will take longer than a piece of silver. Also, firing at a higher temperature will take less time than firing at a lower temperature. Firing transparent enamels directly onto copper takes longer to get to the stage where no reddish oxides show underneath, than when fired on previously fused enamel.

Photo #3

Enamelists have differing firing styles. Some artists like to fire at a higher temperature and some at a lower heat. An enamelist must experiment to determine which techniques work best with the tools and materials available.

Fusing of the enamel can be viewed in various stages as follows:

Description

"Sugar" (just starting to melt)	See Photo #4
More flow	See Photo #5
"Orange Peel" (almost done)	See Photo #6
Smooth and glossy surface	See Photo #7

Photos by Stephen Lanza

Photo #4   Photo #5   Photo #6   Photo #7

Photo by Jennifer Bauser

Photo #8

Keeping Enamels in Good Condition

Enamel must be kept dry. Store enamels in airtight jars in a dry area. See Photo #8. To help keep them dry, some enamelists place a small packet of silica gel in each jar; however, others disagree with this practice. Enamels can last for many years and still retain their beautiful color—simply store them properly.

Particles of enamel that have become wet, and lumped together, can be reground using a mortar and pestle moistened with distilled water, then grade-sifted.

Grade-sifting and Particle Sizes

The process of grade-sifting enamels is done to separate the particles into various sizes, each of which have their own applications in enameling. Individual sizes can be purchased, or enamels can be purchased and then grade-sifted to get various sizes of the same color. This section describes how the process is done, how to name each size, and what application each particle size is used for.

Grain enamel is denoted by its particle size. The unit of measurement is "mesh size," based on the number of holes per linear inch in a mesh screen used for sifting. For example: 200 mesh or 325 mesh. The most common mesh sizes are 60, 80, 100, 150, 200, 325, and 400 or "fines;" 60 being the largest size and 400 being the dusty leftovers. The most common size is 80 mesh and it is sold by all enamel suppliers.

Before grade-sifting, grain enamel is composed of virtually all the above sizes. To grade-sift, first decide what sizes are needed. One sifting tray will be needed for each size, plus two solid trays. For example, if 80, 100, 150, and 200 mesh particles are needed, then in addition to two solid trays, four sifting trays with 80, 100, 150, and 200 mesh screens will be required. It is fastest to grade-sift by stacking all trays together and sifting all at one time. See Photo #9.

To grade-sift:

1. Make a sandwich of a solid tray on top, various trays, in ascending mesh-number order, the lowest number on the top screen layer, and a solid tray on the bottom.

2. Place grain enamel in the top screened tray, (wear dust mask and use ventilation).

Tip #4: Place a coin in each tray to help move the particles through the screen.

3. Swirl and tap the trays to move the enamel through the screens. What remains in each tray is denoted as -x/+y mesh (the screen size above the enamel is listed on the left and the screen size on which the enamel rests is on the right; for example: 150/+200 mesh).

4. Repeat the sifting processing as many times as necessary. Store in appropriate containers and label with color and particle size. See Photo #10.

Clean sifting trays by rapping them on the edge of a garbage can on one side and then the other. Then use a toothbrush or soft cloth on each side. DO NOT use water to clean sifting trays or the enamel will become wet during the next grade-sifting. However, if multiple sets of screens are available, then washing screens between color siftings will keep enamels cleaner.

How should these particle sizes be labeled? It is common to say 80 mesh, but this does not tell if all the sizes are included or only a sifted 80 mesh. To avoid this kind of confusion, the following table describes mesh sizes.

Photo #9

Photo #10

Description	Denotation	Definition
Those that have not gone through any screen.	+100	All particle sizes that sit on top of a 100 mesh screen.
Those that have gone through one screen, but not another.	-150/+200	All particle sizes 150 and smaller (those that fell through a 150-mesh screen) and those that are larger than 200 (those that sit on the 200 mesh screen).
Those that have gone through all screens used.	-200	All particles falling through a 200 mesh screen.
	400 (fines)	Equivalent to -325 (all that falls through a 325 mesh screen). A fine can be defined as the smallest particle of enamel that still has the molecular integrity of that enamel.

When discussing uses of particle sizes, one must compare against sizes of the same enamel color or at least of enamels with the same viscosity. In general, fine-mesh enamel will soften more quickly at a lower temperature than larger particles of the same viscosity; this is useful to remember for hard colors. The finer grains will also pack more densely, thus minimizing "pull-through" over a base coat, and will wet-pack into small or hard-to-reach areas better than larger particles. Finer grains also "bounce" less when being sifted and thus will stay on a surface better without the use of a holding agent. In addition, for any one color, each particle size will have its own tint, hue, clarity, and softening point. Larger particles have a darker, more intense color value. Finer particles are more opaque (less transparent) than their larger meshed equivalents. See Photo #11.

The reason for using larger grains is based on a desire for better transparency. Thus it is not necessary to grade-sift opaques to get the larger particles. See the Limoges —Painting with Enamels project on page 78, which uses different particle sizes. The following is a list of the basic uses of different mesh sizes.

Photo #11

Mesh Size	Uses
-60/+150	Use for surface-tension technique of plique-à-jour. Use over areas that need extra transparent enamel. One example is when you have a design in under glaze that you want to show through an upper layer of transparent enamel.
-60/+325	Wet-packing applications such as for cloisonné, champlevé, etc. Sifting onto liquid enamels.
-100/+150	Use on edges where it will cling a little better than the +100 mesh. Wet-packing into narrow spaces where the use of larger particles may cause pitting or, for opaque yellow-oranges or oranges on copper, may cause black speckling.
-150/+200	Sift on edges to keep them from burning. Sift over domed or shaped metal. Good to help in changing the color of a piece, e.g.: from blue to green.
-150/+325	Good sifting size.
-200/+325	Sift on edges that are really stubborn and hard to cover with enamel.
-200	Helps in creating a wonderful texture when using an underfiring technique such as in limoges enamels.
-325, 400 (fines)	Mixed with holding agent and distilled water, fines can be used as watercolors. They do not create fine lines as with ceramic pigments, but they are lovely.

Cleaning Enamels

To be clear, "cleaning" does not mean that enamels are dirty. They are not. In the past, when grinding their own enamels with a steel mortar and pestle, enamelists put a drop of acid into the enamel to remove iron particles induced by the process of using steel. However, today, steel is not used and thus the iron "dirt" is not an issue. This reference to "cleaning" is "removing the fines."

A freshly purchased packet of grain enamel is composed of different particle sizes. The smaller particles cause transparents to fire cloudy. The fines can be removed by "washing" the enamel; but by grade-sifting first, it will not be necessary to wash the enamel as much. Removing the fines allows the transparent colors to be as clear as possible without speckles in them. The clarity of color is one of the ultimate goals in most enameling techniques.

Some enamelists find no need to wash enamels if being used dry, i.e.: sift; however, others disagree with that notion. Most would agree, though, that opaques need not be washed when used dry.

Some enamelists only wash enamels that will be used for wet-packing. Some only wash them before use. Some wash all their enamels at one time and then dry them thoroughly for later use, either for wet or dry application. This is an issue that each individual enamelist determines for him- or herself.

Photo by Stephen Lanza

Photo #12

Enamels can be washed with household water, if the water is clean and free of harsh chemicals. Otherwise use distilled water. One container is needed to hold the cleaned enamel, and another to pour the wash water into. In general, it is not a good idea to pour the wash water into a drain as the enamel will get into the septic/sewer system and eventually end up in the surrounding water supply.

Tip #5: A good tool to use is a coffee filter sitting in a holder, which will stop any enamel particles from going down the drain. In addition, the enamel left in the filter can be dried and saved for later use as counter enamel.

To wash enamels:

1. Place the amount of enamel needed in an appropriate-sized container.

2. Add some water and agitate the contents. If washing a small amount, the force of putting the water into the container may be enough. If washing a medium amount, something may be needed to stir the enamel. Be certain the container and stirring stick are cleaned between colors. If washing a large amount of enamel, the container may be covered and shaken.

3. Let enamel sit a few seconds.

4. Slowly pour off only the wash water into the coffee filter placed over a clean jar. See Photo #12. Take care not to pour out enamel.

5. Repeat process until the water poured in the jar is clear. If enamel is grade-sifted first, there will be fewer washings than if enamel is not grade-sifted. After grade-sifting, the enamel is usually washed only 2–3 times.

23

| color directly on the metal |
| color over flux |
| color over silver foil |
| color over gold foil |

Figure A

Photo by Stephen Lanza

Photo #13

Robert Kulicke

Photo #14

Test-firing Colors

Transparent and opalescent colors look different on various metals as the metal color is reflected back through the enamel. Also, each enamel is either hard-, medium-, or soft-fusing. See About Enamel Colors below. Be certain to test-fire colors on the selected metal so adjustments can be made to meet reasonable expectations. Also, test-fire layered color combinations to eliminate guess work.

Ceramic pigments fire differently, depending on what they are mixed with. See the Limoges—Painting with Enamels project on page 78 for tips on test-firing ceramic pigments. This section deals with grain-enamel testing.

Test-firing can be done on a mica sheet. This will not show the true color of a transparent when it is on a particular metal; however, it is good for opaques. An instructive exercise is to make a "tab" for each transparent enamel color and run a series of tests. Make a tab, with four sections on the metal used most often. See Figure A, Drill a hole in the top to hang test tabs from a display board. It is possible to use very thin metal (.010"/30ga) for the tests. Be certain to counter-enamel the tabs. Using an overglaze pen, write the color number on the tab. The difference between copper and silver can be easily seen. Copper is on the right. See Photo #13.

About Enamel Colors

Enamels are unlike paint in that they cannot always be mixed to produce other colors. Transparents can be mixed together to create new colors, and transparents and opalescents can be mixed. However, if opaques are mixed together, they will produce a speckling effect, somewhat like mixing multiple colors of sand. In general, this speckling effect is not wanted, but there are some cases when it can be used. One of these is well represented by the "Kulicke Pear" made famous by Robert Kulicke at the Kulicke-Stark Academy of Jewelry Arts. See Photo #14. A pear is a speckled object, and the combination of various opaque enamels produces just the right color. This is also good for stone-type effects. An exception to mixing opaques is to combine the fines of colors that are close in value. When mixing enamels, remember to test-fire before using the enamel mixture in a final piece.

Some colors are "harder" than others—the harder a color, the longer it takes to fuse. Soft-fusing flux may be added to harder colors to reduce their firing time, or the color may be grade-sifted to a finer particle size. When mixing hard and soft colors in the same piece, fire the harder colors early, leaving the softer colors until the end. When unsure if a color is hard, medium, or soft, test-fire it along with a color whose hardness is known. Keep notes on the findings.

Darker transparents can be used over lighter ones, but it is difficult to see lighter colors over darker ones. This effect is similar to washes in watercolors. Use a darker transparent over a lighter one, and sgraffito the unfired darker color down to the lighter color to get a streaked look, like that of hair. See the Sgraffito Plate project on page 118. Transparents can be shaded together and/or used to shade over opaques and opalescents. See Wet-packing on page 32 for blending and shading colors.

Flux can be added to a transparent enamel to lighten the value or hue of the color. Using this method, more layers can be fired without having to make the top layers all flux. Bringing the color to the top of the enamel height has a different look than placing the color low and filling the top with soft flux (this appears as though looking through a transparent window). Both are pleasing, but they are different.

Tip #6: In overlaying flux on a color, one should use soft flux. Flux of any kind can cloud over a color, especially dark colors. Using soft flux or a light transparent will minimize cloudiness. Use thin layers.

Be careful not to overfire opaques—they may turn dark and look burnt on the edges. However, a nice effect can be produced by first firing an opaque color, then firing a different color on top. Try overfiring—the opaque may bubble up through the top layer and give a pleasing random pattern (called pull-through or break-through). This can happen either because of the thinness of the top layer or because of the expansion/contraction properties of the colors used and the temperature at which they were fired. See Photo #15. This may happen by mistake, or it can be made to happen; either way, the results can be very nice.

Tip #7: Warm colors such as reds, pinks, oranges, yellows, lavenders, and opalescent white do not always fire correctly directly onto silver. Fire these colors over a fused layer of flux. Do not fire at too high a temperature. Without the layer of flux, reds will turn brown and the opalescent white will turn yellow.

An interesting gauzy effect can be obtained by first fusing on a color and then, on top, using 1:1 opalescent white/flux mixture. See Photo #16. Add as many layers as necessary to get the exact gauzy look desired. But be careful—one layer too many and the color underneath will be lost.

Photo #15

Ora Kuller

Computer-aided Designs

There are times when hand-drawing a design will not suffice. In these cases, draw the image on a computer. There are many good computer drawing programs available. Some reasons to draw on a computer include:

• The design has symmetric parts that are difficult to draw by hand.

• A photographic image needs to be reproduced.

• A resist process requires a sharp-edged drawing and/or a laser printer output.

• A complex image is needed in a small size. In this case, draw the image large, then reduce and print at the required size. Reduction or enlargement may be done on a copy machine; however, the image from a photocopier may be slightly distorted, which may or may not matter.

• The design may be reused in different sizes, e.g.: make a light-switch plate design that has a single toggle, then make the same design, that has a double toggle.

Photo #16

Karen L. Cohen   Photo by Jack De Geus

• Multiple copies are needed of the same complex design, and using tracing paper would be a tedious chore.

To draw on the computer, one can start with a hand-drawing, scan it, then bring it into the drawing program and trace. Or, simply draw directly in the program.

Metal Patterning for Basse Taille

There are several ways to get a pattern onto metal. Examples are engraving with a hand graver or vibrating pen, etching a design with various types of resists (see the Champlevé Panel project on page 48 or the Basse Taille project on page 42), roll-printing onto the metal (see the Fusion Inlay Under Enamel project on page 62), guilloche, or repoussé. Some of these can be combined as when etching a design onto brass or copper, then using the etched plate to roll-print onto a piece of fine silver.

Roll-printing is done by using a rolling mill to "press" a pattern onto a piece of metal. See Photo #17. The pattern can come from a wide variety of sources. Some examples are textured paper, paper or stencil sheet in which a design is cut, wire formed into a design, textured fabric such as eyelet, and an etched metal plate. Be certain the pattern is deep enough to show under a few layers of enamel.

One must be careful of the design's direction, especially when using letters. That is, when roll-printing a design, the plate must have the design in reverse so when it is roll-printed onto the final plate, the design will be facing the proper direction.

Steps to roll-print onto metal:

1. First anneal the metal in a kiln or with a torch; the softer the metal, the easier it is to roll-print. If there is firescale, then pickle, rinse, and thoroughly dry the metal. If any moisture is left on the metal, the rolling mill may be damaged.

2. Secure the design on the metal with masking tape. Set the rolling mill to have an opening that is just a little larger than the thickness of the metal. Hold the layers tightly together and roll through the mill with one even pass. This may take some practice, but the results are well worth the effort.

Tip #8: If the design is in wire, instead of a rolling mill, a hammer can be used to impress the design into the metal. Hammering requires a slightly crowned, flattish hammer similar to a planishing hammer or the flat face on a standard ball-peen hammer. Hammering should be even so that the wires are pushed evenly into the surface.

Soldering and Fusing

Solder comes in various grades, each melting at a different temperature. Typically, a metal worker will change the grade of solder for each joint made, from hardest to softest, so that the previous joints will not melt when the current joint is being heated. However, for enameled pieces use IT® solder only. It is possible that extra-hard, hard, and other grades of solder can be successfully used, but there is a risk that the softer solders could melt in the furnace.

Photo by Stephen Lanza

Photo #17

It is also possible to fuse the metal instead of soldering. For example, it is useful to first fuse a bezel and then solder it to a back plate; this gives you only one solder joint instead of two. Dee Fontans describes steps in fusing:

1. Cut the fine-silver bezel wire with a small pair of scissors to the correct length. The ends of the bezel need to meet with no gaps.

2. Place the bezel on a firing brick with the seam facing the torch.

3. Without using any soldering flux, flutter the flame in front of the bezel. See Photo #18. The silver will begin to turn orange from the radiant heat. Flutter the flame back and forth very quickly over the seam until the metal melts, or "fuses" together. Remove the heat immediately. (This may need a few practice tries.)

4. There may be some slumping of the bezel wall. If so, file it.

Enameling Surface

In all techniques, except plique-à-jour, vitreous enameling is done on a metal base called the enameling surface, which may or may not be enclosed with a bezel.

Tip #9: Champlevé pieces have integral "bezels" that are the walls of the etching. Larger pieces are almost always made without a bezel; however, jewelry pieces, called jewels, may have a bezel. It is easier to start a piece without a bezel, but the bezel helps secure the edges and reduce warpage. In any case, the enameling surface can be flat or domed, which will also help to reduce warpage. There is no real "correct" way, just advantages of one style over another.

If a bezel is not used, use at least 20ga metal for smaller pieces. If a bezel is used, a metal back plate of 30ga metal is suggested. Very small pieces can use even thinner material. The larger the piece, the thicker the metal base should be.

To dome a piece with a bezel cup soldered on, place the piece on a flat, hard surface, upside down (bezel down). Using first a finger and then, possibly a dap tool or domed hammer, push the surface into the bezel to form a small dome. It need not be high, but doming the enameling surface will help light reflect at different angles and enhance the final appearance. See Photo #19.

Preparing Metal for Enameling

Enamel will not adhere properly to a greasy or dirty surface. It is necessary to clean the metal before beginning the enameling process. Also, if a metal is used such as copper or sterling silver that oxidizes in heat, it will need to be cleaned between firings. Information was collected from each project artist in this book, and it is amazing to see how many different ways there are to clean copper. This section provides general information on cleaning. Use these instructions to clean the metal for the projects in this book.

Usually, cleaning with water and powdered kitchen cleanser will remove any grease and oxidation on the surface of the metal.

Photo #18

Photo #19

To test if the metal is clean enough, see if water "sheets" on the metal surface. Water "sheets" when poured onto metal, then forms an even coating over the surface. If it doesn't sheet, then it will be noticeably pull away from the metal.

Tip #10: If the water is not sheeting, especially on a curved surface, scrub with one drop of liquid detergent and water or try rubbing the piece with pumice powder (can be purchased at a hardware store). Test for sheeting again.

Always clean both sides of the piece because the counter enamel side also must sheet. In addition, be certain to file the edges of the metal to remove any burrs.

Optional cleaning method:

• Heat copper to 800°F, when a visible "oil slick of colors" will move across the surface. This will burn off grease or oil. Heating further will produce copper oxide and make the metal soft, which may or may not be appropriate to project requirements.

• Cool copper and place into pickle solution for 1–5 minutes.

• Rinse in clear water, or an acid-neutralizing mixture of one teaspoon baking soda and one cup water and then in clear water. If the copper object is enclosed, such as a bead, then always soak in baking soda and water before rinsing.

Tip #11: Pickle of any type is a mild acid, so take precautions. In addition, only use copper tongs when removing pieces from pickle so as not to contaminate the solution.

Homemade vinegar and salt pickle solution recipe:

• While wearing rubber gloves, mix white vinegar and kosher salt. Saturate the vinegar with the salt, approximately 3:1 ratio of vinegar/salt. Dissolve as much salt in the vinegar as possible. When salt stops dissolving, the vinegar is saturated.

• Scrub copper with powdered kitchen cleanser, baking soda, or pumice powder with a soapless scrubbing pad.

• Fine silver can also be heated in the kiln to help burn off impurities.

• Scrub fine silver with a soapless glass brush. Be certain to scrub under water so glass fibers are flushed away. See Photo #20.

• Check metal for sheeting. See Photo #21. This is what it looks like when the water does not sheet. Reclean metal. Immediately dry metal with a lint-free cloth to prevent water spotting, which can show under transparent enamel. After cleaning, always hold metal by edges so oil from fingers is not transferred onto the metal.

Cleaning Between Firings

When enameling on copper and sterling silver, firescale (oxidation) will result on any nonenameled metal surface. This firescale may be incorporated into the design, but in many instances will need to be removed. See Photo #22. Some common methods to remove firescale:

• Use a toothbrush and brush off as much as possible, then place the piece in pickle solution for a few minutes, then thoroughly rinse in water.

Photo #20

Photo #21

Photo #22

• To clean tiny areas, use pickle solution with a wire brush or emery cloth.

• If firescale is very thick, try removing it with a wire wheel on a polishing motor then dip in pickle solution to remove any metal burn from the surface. Rinse.

• Be certain metal has a consistent appearance otherwise it will not look uniform with a transparent enamel covering it.

• If using an opaque enamel and if the firescale is very thin, only the loose pieces need be removed. Just enamel over the rest.

• If only the edges of the piece have oxidized, use a metal file or an alundum stone to remove the oxidation.

• To ensure that all particles have been removed from the cleaning process, scrub the piece with a glass brush under running water.

Tip #12: Fine silver does not oxidize in heat and need not be cleaned between firings.

Wirework

• Cloisonné and filigree plique-à-jour require the bending of wires to form the design. Although the two forms of enameling are different, the bending of the wires is similar, as are the tools. See Photo #23. Shown are: round-nosed and chain-nosed pliers, small metal cutter scissors, straight needle-nosed tweezers (#5), and a paint tube crimping tool (optional). Excellent quality tweezers are a must.

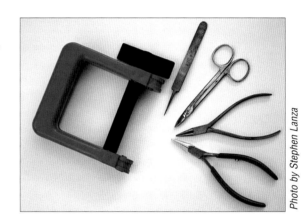

Photo #23

Photo by Stephen Lanza

• For flat pieces, it is convenient to lay out all wires before attaching them to the piece. Make a copy of the design and cover it with double-sided tape. As wire is formed, place it on the tape. This secures the wires until they are needed and allows the artist to see the design unfold. This also allows wires to be easily carried.

• Cloisonné wire is typically .010" thick (30ga), but some enamelists prefer to mill the wire down to .005" in a rolling mill.

Tip #13: Milling the wire thinner will make it longer and wider. It is also possible to hammer down round or square wire to get different thicknesses within one wire. See the Vessel Forms project on page 148.

• Wire for filigree plique-à-jour can be anything from 32ga to any thickness within reason. Before enameling, these wires are soldered together using IT solder to form a framework of metal in which to inlay enamel.

• Before starting, wire may be annealed. Soft wire will not spring out of shape when heated, as hard wire might do. Some enamelists feel that hard wire is easier to work with and thus do not anneal it first. Annealing can be done with a torch, but many enamelists think it best done in the kiln because there is less chance of melting the wire. To anneal in a kiln, make a coil and place on a clean wire screen (one that has never had enamel drip onto it). Place in a 1400°F–1500°F kiln; remove silver when it becomes soft. Remove copper when it turns black. Because

1500°F is below the melting point of either metal, there is no need to worry. If annealing copper wire, be certain to clean off firescale.

• To straighten wire, place one end in a vise; hold the other end with pliers and pull. Or, use two pliers and pull. Pliers can be used to gently "smash" kinks flat. Or, fingernails can be used to run over the kink. If the wire becomes hard, anneal it again; metal tends to harden when it's worked.

Bending Wire

Hold wire with a light touch, it is very thin. Many enamelists do their wirework while holding the wire to the design. This is done on a copy of the design that does not have tape on it. To form the shapes, hold the wire to the design and using an appropriate tool, bend wire as needed. This takes practice to do well.

• To form wide sweeping curves, hold the edge of the wire with tweezers, place wire onto the design and pull, then push and bend the wire into shape. When it is approximately correct, cut a wire a little longer than the proper length. Continue to shape with the tweezers and fingers. When satisfied with the shape, cut to size.

• To make a small circle, cut a length of wire and wrap around the round-nosed pliers until the ends meet. Because the pliers are tapered, reverse the circle on the pliers and shape again. Wire can also be wrapped around a cylindrical object.

• To make a spiral, take a length of wire and wrap it around the round-nosed pliers. But instead of having the ends meets, as for a circle, keep wrapping the wire around itself until it is all wrapped. Remove from the pliers. Using tweezers, while gently holding the coiled wire with the fingers, pull the spiral into shape by separating each layer from the rest of the coil.

• To make a teardrop, cut a length of wire and wrap the middle around the inside of the round-nosed pliers. Push the ends together until they meet in a point. If the ends won't stay together, place the shape on the table and using tweezers, gently squeeze the rounded section until the ends meet.

• To make wavy lines, simply run a length of wire through a paint tube crimping tool. This will produce a fairly even wavy line. Or, wrap the wire around the pegs of a child's pot holder loom. See Photos #24 and #25.

• To make an irregularly crimped wire, use tweezers to make small bends.

Cloisonné Wires

To make straight lines for cloisonné, on a back plate without an outside bezel, make the wire extend past the edge of the piece, then bend it back to rest on the edge, or use this extra to make another straight line on the design. After a few firings with enamel, the extended wire can be cut away.

Alternatively, cut a small slit on the bottom of the wire and bend the two lower parts of the wire in different directions so that they rest on the base of the bent piece. See Figure B. This will help anchor it so it will not fall.

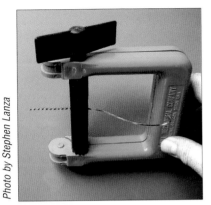

Photo #24

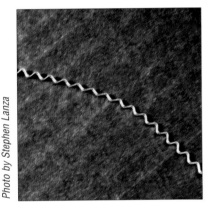

Photo #25

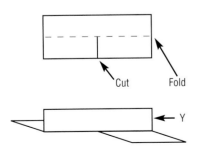

Cut   Fold

Y

Figure B

Tip #14: This will cause the wire to have less height, so use a taller piece of cloisonné wire to match the other wires in the design. That is, in Figure B, the resulting height of "y" should be the height of the rest of the cloisonné wires.

When all wires have been formed, fuse them onto the piece. This can be done one of two ways:

(A) First, fire flux or a colored enamel onto the metal surface. Holding a wire with the tweezers, dip into a small amount of liquid holding agent, then place on the background of fused enamel. Let dry and fire until the wires are set (slightly sunken into the fused enamel). See Photo #26.

Tip #15: If there is trouble telling if the wires have set, try putting a pinch of appropriate enamel into an area and fire. When the enamel is fused, the wires will be set.

(B) If a background of fused enamel is not desirable, wires can still be set by placing, and then wet-packing appropriate enamel against each wire. When the enamel fuses, the wires will be set. The disadvantage of this method is that the wires may not be fully tacked down and some enamel may move from one space to another during subsequent firing.

Photo #26

Applying Enamels

There are two basic methods for applying grain-form enamels—sifting and wet-packing. Remember, not all enamels are compatible so be careful how layers are built up. In general, leaded enamels can be layered over fused unleaded enamels, but not vice versa.

Sifting

Sifting is a method of applying enamels dry; the enamels can be grade-sifted and/or washed and dried first. Factors that affect sifting are the skill of the enameler, enamel mesh size, size of the screen in the sifter, amount of enamel inside sifter, the size and shape of the surface being covered, whether a holding agent is used, the desired pattern, and the evenness and thickness of enamel applied.

Photo #27

Before sifting, prepare the work area by cleaning a space in a ventilated area and getting some clean paper for liners. If a clean paper is used for each color that is sifted, the enamel that falls on this paper can be poured back into the jar. If colors are mixed, use droppings for counter enamel. Wear a dust mask.

Sifters come in a variety of sizes with a variety of screen sizes. See Photo #27. One way to make a sifter is to attach mesh cloth to an embroidery hoop. Use a sifter with a larger mesh size than the enamel being used so that the enamel powder falls through easily. For example, if sifting 80 mesh enamel, use a 40 or 60 mesh sifter.

When sifting, try to get an evenness throughout the layer. Sift in a pattern, at a 90° angle to the surface being covered. The sifting pattern can be a circle, a spiral,

Photo supplied by Jean Tudor

Photo #28

Photo by Stephen Lanza

Photo #29

or in lines. Each path should overlap the previous path by about half. If the surface to cover is curved, hold the piece and tilt to give the 90° angle needed for proper sifting. See Photo #28. If the surface is bumpy, as you sift, rotate the piece to maintain the 90° angle, otherwise clumping will form, which produces an uneven coat. It is better to sift three light coats rather than one thick coat.

Tip #16: To get an even coat of enamel on a flat metal surface, spray the enamel with water until quite wet, sift, then tap the edges with a small object, such as a pencil, to level the enamel layer.

If the piece is flat, rest it on a "sifting holder" which can be a jar top, a stack of coins, or a large-holed screen folded into a zigzag. See Photo #27 page 31. A jar top or stack of coins raises the piece from the flush surface, but keep it low to the work area so enamel isn't flying all over.

Holding agents can be applied, either sprayed or brushed on, before sifting begins. Try a 1:1 Klyr-fire/water solution or use enameling oil. Be certain the coating is as even as possible and use as little holding agent as possible. Before applying a holding agent, have the enamel ready and waiting in the sifter. This will keep the holding agent from drying before the enamel is applied. Before firing, be certain the holding agent is completely dry.

When spraying a holding agent, make a spray box to catch the overspray. This can be as simple as turning a cardboard box onto its side and holding the piece inside the box while spraying.

To hold a small piece while sifting, make a "handle" with masking tape. Take a long piece and fold in half making a "T" shape. Attach the open ends to the enamel and "hold" with the doubled part. See Photo #29. Be certain to remove this before firing!

Clean sifters between colors (see To grade-sift on page 21).

Wet-packing

Wet-packing with grain-form enamel is used in many different enameling techniques. In general, it would be used when applying enamel to an enclosed space or

Photo by Stephen Lanza

Photo #30

for forming a particular shape without a stencil. That is, use when sifting is not practical. Other terms for wet-packing are wet-inlaying and wet-charging. Wet-packing can be done with a paintbrush or a spatula.

Select the correctly sized container to hold wet enamel, and make certain it has a lid so that dust does not get into washed enamels. For small amounts of enamel, use small paint containers from a craft store, another idea is to use small sauce containers from carry-out restaurants or film containers. See Photo #30. Label each container with the appropriate color.

Always wash, and possibly grade-sift, enamels before wet-packing. See Cleaning Enamels on page 23.

To wet-pack, have handy the washed enamels, the inlay tools (paintbrush[s] or spatula[s]), clean water in a small container or two, and paper towels, tissue, or cotton swabs.

After a color is inlaid, use a rolled or folded piece of paper towel, a tissue or a cotton swab to wick out as much of the water as possible. By doing this, the enamel is pulled down to the base, eliminating air pockets. See Photo #31. If the enamel completely air-dries, these air pockets can remain and cause bubbles in the fused enamel. When all the packing is done for one layer, allow the piece to dry before placing in the furnace. If the enamel is not dry, the rising steam from the moisture may cause grains to jump from one place to another. To help eliminate an opaque enamel grain from jumping, add a small amount of holding agent to the washed enamel. Drying time can be reduced by placing the piece under a heat lamp or on top of the furnace.

Tip #17: When the enamel is dry, move the piece carefully because the enamel can shift out of place. To keep enamel from shifting, spray the wet enamel with a 1:1 holding agent/water solution to help set the enamel so that it does not shift as easily.

Tip #18: When drying enamel with the use of heat (on the kiln or under a lamp) be aware that the metal tends to get hot and can burn unprotected fingers.

When working with wet enamel, it can become too dry to handle well. Add a small amount of clean water. Convenient ways to add water are either with a spray bottle or an eyedropper.

Tip #19: Use the trivet as an easel. It can help to tilt the piece for easier access, and when a layer is completed, the piece is already balanced on the trivet.

Keep clean water close at all times. As in watercolor painting, tools frequently need to be rinsed, especially between colors. When the water has too much enamel floating on the surface, change the water. Additionally, change the water after inlaying an opaque color—even a single grain of an opaque enamel will show in a transparent area. See Photo #32.

Tip #20: Once a piece is wet-packed and dried, it is inadvisable to add more enamel until after fusing that layer. However, if absolutely necessary, try wetting the area again, from an edge, and then it may be possible to adjust the enamel. Be certain to dry again.

Helpful Techniques for Wet-packing

• Hold enamel container in the nondominant hand and slightly tilt it so there is easy access to the mixture; the wettest part will fall to the lowest edge. Hold the container close to the area to wet-pack.

• For larger inlay areas, use the enamel that is on the fringe of the wettest part. Pick up the enamel and place it in the appropriate area. Add water to help spread the enamel to the corners and edges, but don't make the layer too thin. If doing cloisonné, do not allow the enamel to spread up the sides of the wire; this can cause the wires to bend inward. Also, if making a background of an opaque enamel

Photo #31

Photo #32

Photo by Stephen Lanza

Photo by Stephen Lanza

that will be shaded with a transparent, if the opaque is high on the wires, it may show as a thin line when finished. Remember to wick-dry the enamel once all the area is packed.

• Small areas require a different technique. Using a paintbrush, add a little water to the small area. Again using paintbrush, pick up some of the dryer enamel from the container and wipe it against the inside edge of the container. Clean the paintbrush in water and wipe dry on a paper towel. Pick up the enamel just wiped on the inside edge and see if it looks fairly dry. If not, repeat: wipe on the edge again, clean the paintbrush, dry on paper towel, and pick up the enamel again. When it is fairly dry, touch the dryish enamel to the water in the small area—the water will pull the enamel right in. Wick-dry as usual.

Tip #21: Change paper towels often. Do not use a piece where a dirty paintbrush has been wiped, as enamel colors might transfer.

• When wet-packing opaques, make the layer as thick as reasonably possible to reduce the number of firings. For opalescents and transparents, it is recommended to make the layers thin.

Photo #33

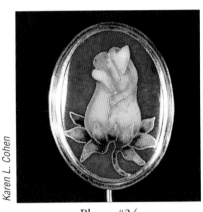

Photo #34

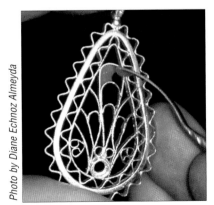

Photo #35

• Two transparents can be blended as with watercolors. To do this, first wet-pack two or more colors next to each other. Add a little extra water by dipping the paintbrush in clean water and touching it to the enamels. Then, using the paintbrush, smudge the "line" between the two colors. By doing this on more than one layer, a pleasing mix from one color to the next is achieved. See Photo #33.

• One transparent can also be used for shading from light to dark. Shading with a single color can be done in two ways:

(A) Use multiple layers, but in a progressively smaller area on each layer. This method takes into account that more/thicker layers of a transparent color will be darker than fewer/thinner layers. To do this, wet-pack transparent color to the widest area that needs to be shaded. Fire, and cool. In the next layer, apply the same color, in a smaller area. Fire, and cool. Continue adding layers with smaller areas until the desired shading is achieved. Fill the remaining space with thin layers of flux.

Tip #22: Do not apply the flux until the shading is complete, thus allowing for adjustments necessary. See Photo #34.

(B) To shade with particle sizes, use the smaller sizes in the darker areas of the shading and the larger particles in the lighter areas. It is possible to wet-pack these various particle sizes all at the same time and thus have fewer firings than with the previous method. Fill in the remaining space with soft flux in thin layers.

• Plique-à-jour enameling (pierced and filigree methods) uses wet enamel, but it is inlaid differently, using a surface-tension method. See Photo #35. Also see the Plique-à-jour Pierced-heart Pendant project on page 102.

Liquid Enamel

Liquid enamels can be applied by pouring, dipping, brushing, or airbrushing. See the Liquid Enamel and Glass Ball Additives project on page 88. Also see the Ginbari Foil Embossing project on page 68 for dipping procedures.

To pour liquid enamel:

1. Make a handle to hold the piece. See Photo 29 on page 32.

2. Mix the enamel to a proper consistency by adding small amounts of distilled water. If too much water is added, simply wait until some evaporates before it is used. To test for proper consistency, dip a clean finger into the liquid, holding it over the jar. The consistency is correct if three drops come off your finger.

3. Hold piece by the handle over a container to catch the runoff. Start at the top and pour enamel over the piece, moving the pouring stream over the top edge. When covered, continue holding over the container and use a clean finger to wipe excess enamel from the bottom edge. Repeat as needed. Let dry. There should be a well-coated smooth sheet of enamel on the metal base. Remove excess from back.

Photo #36

Counter Enamel

Because of stress on the enamel by the metal, when cooling, enamel may tend to crack. To counter the stresses, enamel is also applied on the back of a piece—thus the term "counter enamel." Counter enamel should be the same thickness as the enamel on the front. However, not all pieces may need to be counter enameled. For example, if the metal is thick enough and the enamel is thin enough, then the piece may not need the counter enamel. The need for counter enamel is also determined by factors such as a bezel around the enamel that holds the piece rigid or if the piece is domed. So, how does an enamelist know when to counter enamel? Either choose to always counter-enamel, or simply enamel a piece, but if it cracks, then it must be counter-enameled. See Photo #36.

If the back is not going to show, use scrap enamel for counter-enameling. Scrap enamel can be obtained in many different ways, such as during grade-sifting when one of the particle sizes is not wanted, the enamel left over from the enamel wash water, or when colors have gone "bad" and lose their desired brilliance. Counter enamel can also be purchased from suppliers. Some enamelists never use scrap enamel because they feel the back should look good too, even if it doesn't show.

Counter enamel can be sifted on, or poured or brushed on in liquid form. A combination of brushing on liquid enamel and then dry-sifting enamel onto the wet liquid can also be done. Just be certain to let it completely dry and remove any enamel from findings before firing. Counter enamel can be fired by itself, or let it dry and then enamel the front before firing. Less warpage will occur if both sides are enameled on the first firing.

Tip #23: If doing both sides at once, then the counter enamel should either be done with liquid enamel, or if sifted, thoroughly sprayed with a 1:1 holding agent/water solution before it dries.

Tip #24: If the back of the enamel will not be seen, paint kiln wash or an antiscale solution on the fired counter enamel and dry. This will allow the piece to be fired flat without the trivet sticking to it. The matte surface is also better for use with glue.

Working with Foils

Various types of very thin metal (called foil and leaf) can be used in enameling. See Photo #37. Foil should have holes put in it to allow air to escape during the firing process. The easiest way to do this is by putting the foil on a piece of 220-grit wet/dry sandpaper. Cover with a piece of felt or waxed paper and rub the top with a brayer (a type of roller) or wooden spoon. All pieces can be done, then stored perforated. Label so perforation isn't done twice. The various types of foil and leaf are:

Photo #37

(A) Pure (fine) silver and pure (fine or 24k) gold foil. These are thin enough that they are difficult to hold in the fingers. When cutting foil, it is easiest to hold it between two pieces of paper. The top paper should have a drawing of the design. Hold the sandwiched foil firmly in place and either cut with a small scissors, or with a craft knife.

Foils can be applied to enamels near the bottom layer, in the middle, or on top in a final flash-fired piece. The foil can be made into a solid piece or used in a mosaic pattern. In any case, foil must be adhered to fused enamel. To make certain foil sticks, brush some holding agent onto the fused enamel. Pick the foil up with a paintbrush and lay in place. Be careful not to allow the foil to ride up the side of any cloisonné wires, if present, as this may look like an error when the piece is done. Dry, then fire.

To cut foil to fit a cloisonné cell, take a piece of paper and place on top of the wires. Use a pencil to lightly rub onto the paper to reveal the pattern of the wires. Sandwich the foil between the rubbing and another piece of paper. Cut to shape. Apply to enamel piece as previously described.

Photo #38

To cut foil to use in a mosaic pattern, use tweezers and fingers to rip foil into small pieces. This is a good application to use bits and pieces from earlier cuttings. Apply onto fused enamel as previously described. The pieces can either touch or not touch, depending on the design. See Photo #38. If this method is used on the third layer of a cloisonné piece, it can give the enamel a crinkled look and help the colors reflect at different angles. See Cloisonné Brooch project on page 52.

(B) Leaf is very thin and cannot be held by hand or it will fall apart. Patent leaf comes with a backing, making it easier to handle. Leaf comes in gold, silver, and palladium.

To apply leaf:

1. Wipe fused enamel surface with alcohol to clean and remove grease.

2. Paint an area with enameling oil and dry to tacky stage (about one minute under a heat lamp).

3. Place leaf onto the oiled area. Let dry completely (two minutes or more under a heat lamp).

4. Using a soft brush, wipe away the excess leaf. More leaf can be removed with a swab moistened in alcohol. To distress leaf, rub lightly with #000 steel wool.

5. Fire and cool. Enamel can be applied over leaf, if desired.

(C) Ginbari foil is a thicker silver foil that is easier to handle. See the Ginbari Foil Embossing project on page 68.

Photo #39

Ginbari scraps can be chopped in a blender with water. They can then be sifted into sizes (small, medium, and large) and used with stencils to make surface designs. To apply, paint an area with enameling oil (a stencil can be used to shape foil design), dry until tacky, then sprinkle with chopped foil. Fire and cool.

Tip #25: When using silver foil or leaf, remember that some colors change when directly applied to silver. Cover with one layer of flux first. See About Enamel Colors on page 24.

Tip #26: Silver foil and leaf can tarnish. Fire at least one thin layer of flux on top of silver foil or leaf.

Finishing

Some enameling methods require the enamel be stoned down to an even surface. This may need to be done during the middle of the enameling process. See the Limoges—Painting with Enamels project on page 78. Or stoning (grinding) can be done at the end of the project. See the Champlevé Panel project on page 48. Finishes can be matte or glossy. Tools are simple and easy to use. See Photo #39.

Photo #40

A matte surface can be achieved by stoning (using abrasives up to 600-grit), underfiring, or using a glass etchant. Each gives its own look. See Photo #40. The purple-and-black area was matted using emery paper and the rest was finished by underfiring, then etching. Underfiring sometimes gives a pleasing bumpy look to the enamel.

A glossy surface can be achieved by flash-firing or hand-polishing, both are preceded by stoning. Of course, if there is no stoning at all, the last firing will be glossy, but not necessarily smooth. In cloisonné particularly, it is easy to see a difference between flash-firing and hand-polishing. Although both are glossy with a flash-fired finish, it will be possible to feel the cloisonné wires; but a hand-polish results in a mirror finish in which the entire top is smooth to the touch. See Photos #41 and #42 for a comparison on the same piece. Photo #41 is flash-fired, Photo #42 is hand-polished.

Photo #41

Photo #42

Photo #43

When stoning and hand-polishing, it may be easier to hold a small piece by attaching it to a dop stick, which is normally a stonecutter's tool. Consult a stonecutting manual for details, but the gist of the technique is to heat the dop wax onto a stick (like a wooden dowel). Heat both the dop wax and the enamel piece, then push the dop onto the enamel back. Mold the dop to the back's shape, then cool. To remove, place the piece in a freezer until it pops off the stick. See Photo #43.

Stoning

Stoning is a way to level out an enamel surface and remove the enamel from the raised metal surfaces. Stoning abrasives come in different grits, like sandpaper, and can be on a motorized mechanism or on a hand tool. The lower the number, the coarser the grit. When stoning, start with a low number and work up to a high number. In the first (course) stoning, the object is to get all the enamel even. In successive (fine) stonings, the goal is to remove the "scratches" made by the previous stoning. Some enamelists, like Don Viehman, will flash-fire the piece after the course-stoning stage and then proceed with fine-stoning and hand-polishing. He feels that the flash-fire provides the best rounded edge, which makes operations such as setting in a bezel easier.

Stoning can be done with wet/dry sandpapers, alundum stones, diamond abrasives (papers or files), or motorized lap wheels with diamond compound.

When stoning, be certain to use plenty of water so that the piece doesn't heat up and the "dirt" is washed away. See Photo #44. Always clean with a glass brush after stoning and before firing.

Tip #27: If stoning on flat wet/dry sandpapers, move the piece back and forth in the same direction and buffer the sandpaper on a few sheets of regular paper. In all methods, try to tilt the piece to achieve a rounded surface. In some cases, as when doing champlevé, the metal will also get stoned and polished.

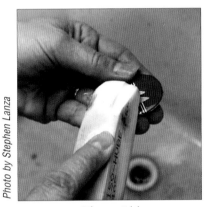

Photo #44

Of course, not all enamelists agree on the sequence of grits to use for stoning. Two sequences are presented, the first in detail:

1. Start with a 320-grit abrasive for small pieces to even out the enamel (possibly use 220-grit on the edges as these may take longer to do with the 320-grit).

2. Clean and flash-fire if desired. To flash-fire, place piece in a 1500°F–1550°F kiln and fire until just the top surface flows. This can be a final step or more stoning and hand-polishing can be done.

3. Use a 400-grit abrasive next. If the piece has been flash-fired, use the 400-grit to even out the piece again. If the piece was not flash-fired, use the 400-grit to remove the scratches made by the 320-grit.

4. Clean, then use a 600-grit abrasive to remove scratches made by the 400-grit. This can be a final step if a matte finish is desired.

5. Clean, then use a 1500-grit abrasive to remove the scratches made by the 600-grit. Proceed to hand-polishing. This step may be skipped and hand-polishing begun. However, if the 1500-grit step is skipped, the polishing will take longer.

Another sequence of grits is to start with 100-grit, proceed to 220, then 320, then 400, and finally to 600. Leave with a matte finish or continue to hand-polish.

Note: Some enamelists clean pieces using an ultrasonic cleaner with distilled water.

Hand-polishing

Hand-polishing is done with a compound called cerium oxide, which is also used for polishing certain types of stones. This can be applied on a motorized or hand tool. The procedure for this step is basically the same as for stoning, except cerium oxide is put on soft leather with some water. As with stoning, use water to cool the piece. Polish with cerium oxide until a mirror finish is achieved. See Photo #45.

Metal finishing

Many times the metal shows on the final piece. Decide if a high polish or a matte finish is desired on the metal. If the stoning steps above were used with at least 600-grit, the metal can be polished with jeweler's rouge. A matte finish can also be achieved with a glass brush.

Photo #45

On bowl forms or wall pieces that are totally covered with enamel, the edge may be the only metal that shows. This can be finished by filing with a jeweler's file and then polishing with metal polish. Alternatively, Averill B. Shepps suggests using a soap-impregnated steel wool pad designed for kitchen use to give a final finish. The soap keeps the steel wool bits from becoming airborne, and the steel wool will smooth out any stoning marks. Naturally, this method is used at a sink and not near the enameling area, just in case a bit of steel comes loose.

Mounting

Wall pieces need to be mounted. Three projects in this book suggest ways to mount enamels. See Mounting Enamels on page 98. Also see Limoges—Painting with Enamels project on page 78 and the Torch-altered Metal with Cheesecloth Stencil project on page 136.

Miscellaneous

Here are some tips that don't seem to fit in the previous categories.

Tip #28: When firing millefiori, glass lumps, or frit, first prefire on a piece of mica to create an approximate shape. Cool, and clean the mica off the back, then embed as usual. If firing multiple pieces of millefiori, place a few particles of flux between pieces. These pieces must have counter enamel before setting the millefiori. See Photo #46.

Tip #29: Use a "light" table to help draw layers of a design, or when tracing. Small light tables can be purchased at art supply stores.

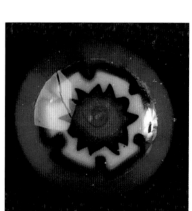

Photo #46

Tip #30: Keep track and make notes of pleasing effects and color combinations.

Tip #31: "To make a very fine line, use a very fine felt-tip pen. Draw your lines and immediately sift a thick coat of opaque over the lines. Tap off excess. Sift flux . . . lightly over the lines and fire. The ink fires away completely. You can also make white or yellow lines on a dark background." (quoted from Assorted Pearls and Gems)

Ingrid M. Regula

INGRID

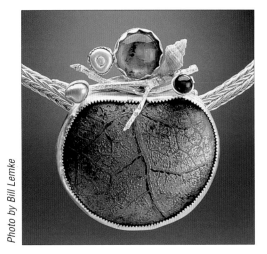

German born, Regula received an art education at the Academy for Fine and Applied Arts, Berlin, before immigrating to Cleveland, Ohio, in the 50s. After getting married, she pursued a career in Graphic Arts, starting at the American Greetings Co. Motherhood changed this path and when her two sons were old enough, she enrolled in classes at the Cleveland Institute of Art inspired by Kenneth Bates, Charles B. Jeffery, and William Harper. She fell in love with the art of enameling, the color and the process. Regula experimented with the different techniques and soon started to compete in local shows and to sell some work in local shops and galleries. Kilns and materials were in easy reach in Cleveland then, and she was able to furnish her first studio.

A move to Milwaukee, Wisconsin, with her family in 1970 took her to a cooler climate as well as to a less active enameling arena. Regula found strong representation of metalsmiths coming out of the University of Wisconsin, but with little or no enameling background. A fine-metals program was offered at the Milwaukee Area Technical College (MATC) where she started to teach enameling at the neighborhood campus. At the same time, she expanded her own abilities by enrolling in art metal classes in order to work three-dimensional forms for enameling. She learned to raise her first 10" bowl and prepared it for basse taille enameling.

Admiring the work of June Schwarz of California, who is a master of the basse taille technique, Regula proceeded to try her luck with it. She learned to tool the metal with hammers and punches. She then laid transparent enamel over these surfaces, fired them, and thus a new depth was created. Etching was the next step to further enrichment. She started to make boxes with wearable lids. The challenge of one-of-a-kind pieces kept things fresh and interesting.

Regula's studio became a place full of parts, gemstones, and beads that were waiting to be put together into just the right way. She gradually worked into the jewelry market. When she was not busy with art fairs or teaching, she traveled to take another enameling or metal class at Arrowmont School of Crafts. She learned to photograph her work, which was most important in order to get accepted into juried shows.

"I established a wonderful relationship with Edgewood Orchard Galleries in Door County, Wisconsin, where my work 'fits' and has been sold now for almost 20 years.

"After 30 years in Wisconsin, another move found me in Florida. Again I have a 'room with a view.' My workshop looks out at a flower garden and lake with water birds and an occasional alligator. I don't do art fairs anymore but my garden presents new challenges. My gardening hobby was always like a golden thread woven through my fabric. Nature provided the spirit and nourishment for my ideas. Color and light expressed in enamels are a stimulus for me, like they would be for a painter. When I step into my studio, I feel excited to know that there are still more enamels to create and something better to be achieved."

Photo by Bill Lemke

Phoenix Rising Basse Taille & Amethyst

Basse Taille Copper Enamel & Sterling with Azurite

Basse Taille Copper & Foil with Tourmaline Beads

Basse Taille

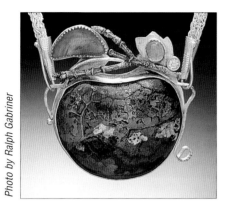

Photo by Ralph Gabriner

TOOLS:

Brass brush
Finishing tools
Firing tools
Flexible shaft with drill press
Glass brush
Glass or plastic etching container
Glass slab
Hydraulic press and associated
    equipment with 20-ton jack
Jeweler's saw, saw blades, and spiral
    blade
Paintbrushes: 0, 00
Pickling tools
Plastic scrubbing pads (2)
Sifters: 80 and 100 mesh
Tongs
Urethane pad: 2" x 3"

MATERIALS:

Acrylic sheet: 2" x 3" x ¼" thick
Ammonia
Asphaltum varnish resist: 1 small can
Baking soda
Contact paper
Copper sheets: 2" x 3", 22ga, (2)
Cotton swabs
Enamels: Thompson lead-free 2020
    clear; 2305 Nile green; 2410 copper
    green; 2325 gem green; 2760 mauve;
    2420 sea green; 2810 geranium pink;
    1420 mint green (for counter
    enamel); black underglaze
Ferric chloride: 1 pint
Fresh large leaf with distinct vein
    pattern
Klyr-fire®
Lint-free cloth
Liquid detergent
Liquid resist
Mineral spirits
Newspapers
Powdered kitchen cleanser
Silver foil
Watermelon tourmaline; 1 green
    tourmaline cab 6mm x 8mm; and one
    CZ, 4mm
White plastic spoons

Techniques to Know

- Applying Enamels: Wet-packing, pg 32
- Cleaning Between Firings, pg 28
- Cleaning Enamels, pg 23
- Finishing, page 37
- Preparing Metal for Enameling, pg 27
- Grade-sifting and Particle Sizes, pg 20
- Working with Foils, pg 36

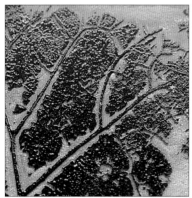

Photo #1

For information on soldering and chain making, refer to Appendix on page 156.

Metalwork:

1. Prepare two 2" x 3" pieces of copper for enameling using the furnace option, then clean copper pieces with powdered kitchen cleanser.

2. On newspaper-covered worktable, spread asphaltum varnish resist on a slab of glass. Press part of fresh leaf on resist, coating it evenly. Print it on newspaper one time to get just the right image, then imprint on two copper sheets, making an extra image to chose from and for earrings. See Photo #1. Use 22ga copper, because it is light enough for jewelry and still deep enough to etch and be effective under transparent enamel colors. Cover back side of copper with contact paper and paint edges with liquid resist. Dry for several hours.

3. Etch copper pieces in ferric chloride in a heavy plastic or glass container. Place face down in solution. The piece can be laid on plastic scrubbing pads. Or tape copper with package sealing tape and attach to random Styrofoam® blocks to float face down in acid. Or, immerse pieces vertically. Nonetching surfaces must be covered completely. A rod can be laid across vat and pieces hung on plastic covered wire for easy removal. Check etching progress carefully. Using new acid, it will take about two hours to etch to proper depth which is approximately ⅓ the way through metal. See Diagram A.

Photo #2

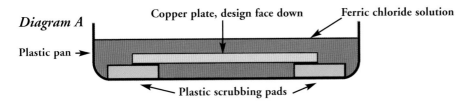

Diagram A

Copper plate, design face down — Ferric chloride solution

Plastic pan →

← Plastic scrubbing pads →

Note: *Refer to etching manual for safe handling, use, and disposal of etching solution.*

4. Using tongs, remove copper plate and drop in bath of baking soda and water. Rinse well. Remove asphaltum with mineral spirits and scrub with soapy water and ammonia to make certain all residues from etching have been removed.

5. Using jewelry saw and spiral saw blade, cut shape of the design from acrylic sheet to make matrix die. See Photo #2. The cutout should be centered and no closer than ¾" from the edge of the die. File and sand inner edge perfectly smooth.

Note: *If a hydraulic press is not available, the copper plate may be cut and domed by other means in order to finish the project.*

6. Anneal etched copper plate in furnace at 1350°F.

7. Isolate area for the design with matrix die on etched copper piece. Mark silhouette and place both on a piece of urethane, sandwich fashion, between platens on the hydraulic press. See Photo #3. Under pressure, the metal stretches into cutout and a pillowed form results with a flat flange. The etched pattern will appear undisturbed. Cut or saw flange away. Utilize extra pieces for tests and earrings.

Photo #3

Enameling:

1. Clean piece for counter-enameling.

2. Using 80 mesh sifter, sift a coat of 1420 mint green on back of piece and fire at 1500°F for two minutes. Apply second coat with 2420 sea green in the center. See Photo #4. Fire again, cool, then place in pickle solution for three minutes. Rinse and dry.

Photo #4

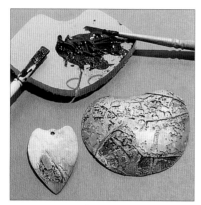

Photo #5

Photo #6

Photo #7

3. Using brass brush, clean front of piece. Prepare small amount of black underglaze on a flat surface and brush it across etching so it accents the lines. See Photo #5. Place piece on top of furnace or under heat lamp to dry. Using soft cloth, wipe over piece in one direction to remove excess black underglaze.

4. Prepare enamels for wet inlay by using a separate plastic spoon for each planned color, rinse enamels with distilled water, apply a drop of Klyr-fire and mark the numbers of colors on the spoons. Use a fine, animal hair paintbrush to inlay the enamel. Start with 2020 clear enamel at top left as an undercoat for transparent red that will be applied in the later stages. See Photo #6.

5. Inlay 2305 Nile green next to the clear, which also will receive 2810 geranium pink later. This makes a gentle transition of red tones. Follow with 2410 copper green and alternate with 2325 gem green and some clear until piece is evenly covered. Tap side of piece with a tool and moisture will float to top to be easily blotted out at the edge with a cotton swab.

6. Allow enamel to fully dry, then fire at 1500°F for two minutes. First firing should be hottest, as it serves to remove cuprous oxide, a layer that initially darkens colors if the temperature is too low. A second 1500°F firing should clear any remaining cuprous oxide. From then on fire at 1450°F. See Photo #7.

7. Using alundum stone or diamond cloth under running water, smooth surface lightly. Clean with glass brush under running water. Brush diluted Klyr-fire over piece and sift 2810 geranium pink on left side corner extending on over light green areas. Blend with 2760 mauve and greens. In center, apply bits of silver foil with Klyr-fire and fire all together at 1450°F for two minutes. See Photo #8

8. Paint small amount of black underglaze on foil and dry. Using 100 mesh sifter, apply thin coat of all colors selectively over entire surface. Fire at 1450°F. Stone again lightly. Clean with glass brush under running water.

Finishing:

1. Sift a thin coat of 2020 clear over entire surface. Fire again for 90 seconds. File edges and mount piece in setting. See Photo #9.

Photo #8

Photo #9

Enameling Photo Gallery
Examples of Fine Enamel Pieces from Professional Enamel Artists

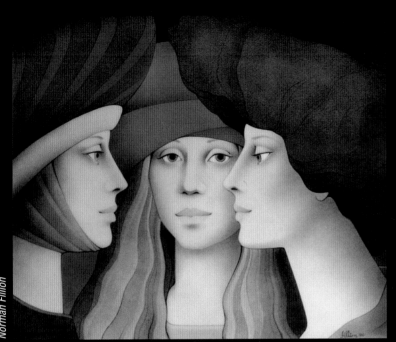

Norman Fillion

Whispers Sifted enamel

Susan Willis   From the collection of Charles and Ashley Wile

Yesterday Fold-formed copper

Barbara N. McFadyen

Basse Taille 24k foil with underglaze and transparent enamels

Christina T. Miller   Photo by Linda Darty

Enameled Brooch

Katharine S. Wood

Although Wood enjoys creating many types of enamels, champlevé is her favorite technique. She feels that there are no artistic limitations with this method, and that anything one can draw, one can etch. Wood, along with colleague Paul Silverman, has been involved in a major development of a revolutionary etching technique for enameling, which utilizes the use of a heat-transfer method traditionally used for printed circuit boards. This technique, now used internationally, allows for low-tech, accurate reproduction of detailed and complex designs for etching.

Wood also loves texture in enamels and often will combine her etching with repoussé, multilevel etches, and embedded components. She feels that, even after more than 15 years of etching and champlevé work, the endless variations keep the process challenging and exciting. Experimentation, not only with champlevé but also with unconventional aspects of enameling (such as firescale), is another major interest.

In the past several years, most of Wood's work has been wall pieces and fluctuates from totally abstract designs to very precise representational subjects such as portraits. Some themes are dominant in her work—jungle/forest/nature imagery, city subjects, or abstract uses of patterns and textures. The human face or form also are frequently incorporated into these themes.

Teaching is an important part of Wood's career. She currently teaches at the Newark Museum in New Jersey, and the 92nd Street Y and the Craft Students League in New York City. Numerous workshops have included sessions with seniors, school children, and youths in the juvenile justice system.

"Teaching is never dull! It keeps you on 'the cutting edge' and often is a source of inspiration. I feel I owe endless thanks to my students, who have enriched my enameling outlook with their creativity and their endless challenges. They have kept me 'open' to all possibilities, which in enameling are endless."

Tribal Rites Champlevé panel 6" x 6"

Chac III Champlevé panel 24" x 24"

Terra Incognita Champlevé panel 6" x 6"

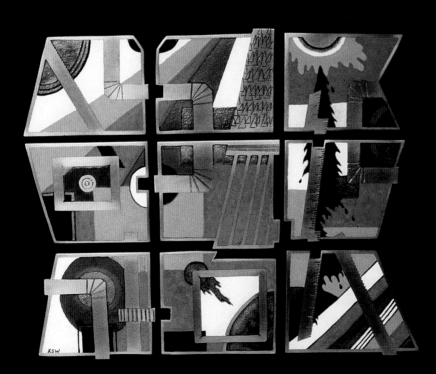

Rocket Machine Shop Champlevé panel 18" x 18"

Champlevé Panel

TOOLS:
Finishing tools
Firing tools
Iron with nonstick surface
Paintbrushes: 0, 000
Photocopier
Plastic pan (at least 4" sq.)
Plastic scrubbing pads (2)
Rubber gloves
Scribe
Sifter: 80 mesh
Wooden block (at least 3" sq.)

MATERIALS:
Acetone
Baking soda or ammonia
Black liquid resist or opaque paint pen
Contact paper
Copper plate: 2" sq., 16–14ga (flat)
Design of choice 2" sq. or use design
   from Photo #1
Enamels: 80 mesh, artist's choice;
   project uses light and medium blues
Ferric chloride
Klyr-fire®
Lint-free cloth
Powdered kitchen cleanser
Printed circuit board transfer film
Rubbing alcohol

Techniques to Know
- Applying Enamels: Wet-packing, pg 32
- Cleaning Between Firings, pg 28
- Cleaning Enamels, pg 23
- Preparing Metal for Enameling, pg 27
- Finishing, pg 37

Metalwork:

1. Photocopy design onto matte side of printed circuit board transfer film. Set photocopier so black areas are solidly black and white areas are clean.

Note: Design should be flopped so any writing is backwards because image will be reversed when transferred onto copper plate.

2. File away any burrs on edges of copper plate. Using plastic scrubbing pad, scrub image area with powdered cleanser until it is clean enough for rinse water to "sheet" off the metal. Using lint-free cloth, dry.

3. Preheat iron on "hot" or "cotton" setting.

4. Set copper plate on wooden block. Using cloth, wipe down image side of copper plate with rubbing alcohol to completely degrease it. Avoid touching image area.

5. Place printed circuit board transfer film with printed (matte) side down on copper plate. Iron with smooth, even motions over shiny side of film until entire design shows through and the design adheres to the metal. Turn wooden block to reach all areas of the film with the iron. Cool.

Note: If the iron is too hot, the film will bubble, if the iron is too cold, the film will not adhere to the copper plate. Adjust temperature as necessary.

6. After film has cooled, peel from copper plate. Design will be transferred onto the plate. Using scribe, scrape off unwanted glitches in the design. See Photo #1.

Photo #1

Author's Note:
I used PNP Blue® transfer paper to transfer the design and block the etch in this project. It is made by Techniks Inc., Ringoes, NJ, and available at various enamel supply companies.

7. Cover back of copper plate with contact paper. Paint side edges of copper with either black liquid resist or opaque paint pen.

8. Wearing rubber gloves, place copper plate face down into ferric chloride solution. Since piece should not touch the bottom of the plastic pan, it can be "bridged" or rested on two clean plastic scrubbing pads. See Diagram A.

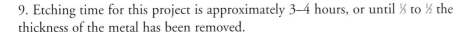

Diagram A

Copper plate, design face down   Ferric chloride solution

Plastic pan →

← Plastic scrubbing pads →

Note: Refer to an etching manual for safe handling, use, and disposal of etching solution.

9. Etching time for this project is approximately 3–4 hours, or until $\frac{1}{3}$ to $\frac{1}{2}$ the thickness of the metal has been removed.

10. When etching is complete, remove copper plate from etching solution, rinse, then neutralize with either straight ammonia or a mixture of baking soda and water. Rinse again. Peel off contact paper. Using cloth and pure acetone, remove circuit board transfer film residue and black liquid resist (opaque paint). Clean entire piece with powdered cleanser and rinse thoroughly. See Photo #2.

Enameling:

1. Clean metal in preparation for enameling. Clean enamels in preparation for wet-packing. Use a drop of Klyr-fire in enamels if using opaque colors.

2. Using paintbrushes, wet-pack first layer of enamel as evenly as possible, leaving room for another 2–4 layers. See Photo #3.

3. Fire piece at 1500°F–1550°F for 1–3 minutes or until enamel is fully melted and glossy. Cool.

4. Clean metal of firescale by pickling and rinsing. See Photo #4

5. Brush on Klyr-fire. Sift counter enamel onto back (unetched) side of piece. Fire, cool, then clean.

6. Repeat Steps 2–4 until enamel is level or slightly above the level of the copper.

Finishing:

1. Use finishing techniques until enamel is smooth and level with the metal and metal design is exposed, remove scratches, then flash-fire to glossy finish.

2. Hand- or machine-polish front and sides of copper plate. Gold-plate if desired.

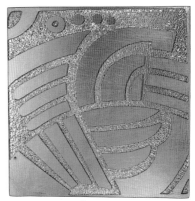

Photo #2

Photo #3

Photo #4

49

Karen L. Cohen

Artwork was always encouraged by my family with outside classes and home projects with my mother and sister. We worked in a variety of media from painting to tilework. However, my love for fine metal and enamels started as a teen at summer camp where we had a silver jewelry activity that I pursued each summer. Being the practical person that I am, at Ohio State University I got a degree in Education and taught high school math for three years before getting my Masters in Computer Science. I then worked as a computer scientist in development at AT&T Bell Laboratories (now Lucent Technologies), where I earned a patent and was published in international journals.

Subsequently, I left the labs to run my own computer consulting company. During all this time, though, I never gave up my artwork and took classes in metalwork and enameling at the School of Visual Arts in New York City, Peters Valley in Layton, New Jersey, and in private classes.

I work primarily in cloisonné, enameling on fine metals for their luminosity, the depth and variety of color, and because of a love of the process. As a computer programmer, I turned complex processing into elegant software. With enamel, I embrace a similar process, as thousands of irregular grains create an object of beauty and depth, one that needs light to bring it to life. I frequently contrast transparent and opaque enamels, and add even more dimension by layering transparents over opaque or opalescent enamels while shading those transparents to create color variations.

My attention to details and to aesthetics means that in all my jewelry, boxes, wall pieces, and sculptures (some with wearable parts), I strive to ensure that every curve is beautiful—one that is as mathematically correct as from a French Curve. The work should be able to be appreciated for the entirety of its emotional quality and for each individual line. Between the lines, the grains, and the layers, I hope to create a constantly shifting resonance with my audience, a resonance they call beauty and feel as joy.

Photo by Ralph Gabriner

Wilsonara Lyoth Ruby Orchid with Chinese Vase Cloisonné panels 13" x 19"

Cloisonné Brooch

Photo by Ralph Gabriner

TOOLS:

Cloisonné wire tools
Containers with lids to hold
   enamels
Finishing tools
Firing tools
Flexible shaft with split mandrel
Jewelry tools
Metal drill bit, #57
Paintbrushes: 10/0, 0; sable
Pickling tools
Small flat brush
Soldering tools

MATERIALS:

Bezel: 4⅓" fine-silver wire –
   or 2mm fine-silver bezel wire
Brooch Background: 2", 20ga fine-
   silver disk
Cloisonné wire: 2', .010" fine-silver
   wire
Enamels: +200 mesh, artist's choice
Fine silver for balls: small amount of
   flat sheet or thin wire or scrap
Gold foil (very little)
IT® solder
Kiln wash (2 tablespoons)
Klyr-fire®
Paper towels
Pin findings:
   Hinge: ³⁄₁₆", 10ga square sterling
   silver wire
   Catch: 1", 16ga round sterling silver
   wire
   Stem: 4", 18ga round sterling silver
   wire, hard
Silver foil: 1" x 2"

Type: denotes if the color is Transparent or Opalescent.
Mixture: denotes how colors were combined for use in this project.

Description	Type	Mixture
medium blue	T	1 part enamel to 1 part flux
dark blue	T	1 part enamel to 1 part flux
light brown	T	Straight enamel, no flux
light brown/yellow	T	1 part light brown to 3 parts yellow to 1 part flux
clear	T	Straight flux
light green	T	Straight enamel, no flux
medium green	T	1 part enamel to 1 part flux
dark green	T	1 part enamel to 1 part flux
light orange	T	1 part enamel to 1 part flux
purple	T	1 part enamel to 2 parts flux
red	T	2 parts enamel to 1 part flux
yellow	T	1 part enamel to 1 part flux
yellow	Op	Straight enamel, no flux

Techniques to Know

• Applying Enamels: Sifting, pg 31
• Applying Enamels: Wet-packing,
   pg 31
• Basic Jewelry Skills
• Cleaning Between Firings, pg 28
• Cleaning Enamels, pg 23
• Counter Enamel, pg 35
• Finishing, pg 37
• Preparing Metal for Enameling,
   pg 27
• Wirework, pg 29

This project introduces cloisonné with an abstract landscape design. Usually cloisonné is done as a "jewel"—a separate piece that is set like a stone into a setting. This project, however, takes a different approach; the cloisonné is done directly into the setting.

Metalwork:

1. To make the cloisonné setting, first make a bezel by taking a 4⅓" length fine-silver wire and make a bezel approximately 1⅜" in diameter (c = d * 3.14). Square up ends of wire for fusing. Anneal as needed.

2. Using torch, fuse ends together. Ends could be closed with solder. However, there will be three other soldering operations so fusing is best. Pickle and rinse.

3. Shape wire into a circle.

4. Using very small solder pieces on inside of bezel, solder bezel to back plate. The inside is the enameling surface so do not get solder on this surface. If solder does show, carefully clean it off with a file so marks won't show under enamel. Pickle and rinse. See Photo #1.

5. Dome enameling surface.

6. Make pin findings:

A. *Hinge:* Drill hole towards top of sterling silver square wire. Round off top edges. Using jewelers saw, cut off approximately ³⁄₁₆" length. File bottom of wire (away from hole) flat.

B. *Catch:* Flatten approximately ³⁄₁₆" of 16ga round wire and bend other end to make pig's tail shape.

C. *Pin:* Fit into hole drilled into hinge. If necessary, enlarge hole. Set aside.

7. Solder on findings. Pickle and rinse. See Photo #2.

8. To make balls which will be used on brooch during last firing, cut small chips of fine silver from flat sheet or wire. Ball blanks may be different sizes, or they can all be the same size. Place silver chips on charcoal block that has never been used with soldering flux. Heat each blank with the torch until it melts into a ball. Cool slightly, remove from block, and repeat process. Make at least 75–80 balls.

Enameling:

1. Review Studio Basics, Tips, and Tricks; Wirework page 29.

2. Refer to Diagram A for design, with each section numbered for later reference.

3. Using glass brush, clean metal by scrubbing under running water. Be certain that water sheets on metal. If ochre used for soldering is difficult to remove, place soldering paste flux on it and heat again. Pickle and rinse. Reclean with glass brush under running water.

4. Counter-enamel a thin layer. Fire and cool.

Note: *Counter enamel (2 layers) will show, so use enamels that complement front of brooch.*

5. Brush small amount of Klyr-fire onto enameling surface and sift small

Photo #1

Photo #2

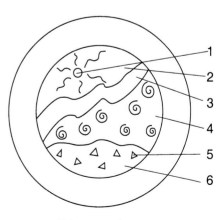

Diagram A

53

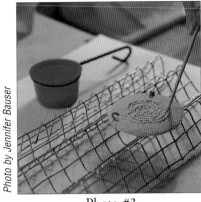

Photo #3

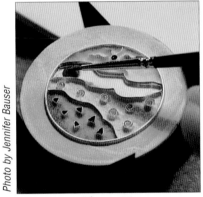

Photo #4

Photo #5

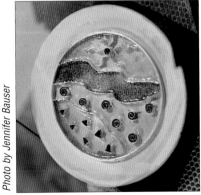

Photo #6

amount of dry 80 mesh flux onto it. Brush away any flux that falls onto outside metal surface. Dry. Fire to glossy stage. Cool.

6. Apply another layer of counter enamel and fire. Use enamel that works with front design and first layer. Using sgraffito or some other technique, make design if desired. See Photo #3.

7. Place small amount of Klyr-fire into shallow container. Bend wires to fit design, dip into Klyr-fire and lay them on flux-fired enameling surface. See Photo #4. Allow Klyr-fire to dry. Fire until wires are set. Cool.

8. Set up enamel colors by placing small amounts (¼–½ teaspoon) into containers with lids. If needed, mix up more later. Flux is added to some colors. The "mixture" used for each color is noted in the beginning of this project.

9. Wash selected enamels.

10. First few layers will not look too exciting because colors will be basic with virtually no shading. Using 10/0 paintbrush, wet-pack sections. See Photo #5. Dry, then fire. Cool. See Photo #6.

Coloring section details:

1. *Section 1:* Two layers of medium blue, then a firing with gold foil. Apply more layers of medium blue until desired color is achieved. Fill in section with flux.

2. *Section 2:* Wet-pack with transparent yellow. On subsequent layers, wet-pack light green around the sun's wires, using transparent yellow for most of the background. Shade with light brown. Fill in section as needed with flux.

3. *Section 3:* Graduate colors from medium green to medium blue to purple, starting at left. Repeat for second layer. Third layer is mosaic pattern in silver foil, which is held with Klyr-fire. See Photo #7 on page 55. Continue with same three colors until depth of color is achieved. Fill in section as needed with flux.

4. *Section 4:* Apply red into spirals and opalescent yellow in all other places. Use opalescent yellow for either one or two layers or until entire background is solid (except around spirals). Shade with red and light orange. The less applied on top of yellow, the lighter the area will be. Fill in section as needed with flux.

5. *Section 5:* Use purple for approximately three layers. Fill in section as needed with flux.

6. *Section 6:* Use light green as base layer. Shade with dark green and dark blue. Fill in section as needed with flux.

7. Enamel at least as high as wires in all areas of brooch. Check enamel level with tweezers to detect any dips. This piece will probably require at least seven layers plus counter enamel, original flux, and wire setting firings.

Finishing:

1. Grind down enamel to make it even with wires. See Photo #8.

2. Using wet/dry sandpaper, clean metal border. This can be done either by hand or by placing wet/dry sandpaper in a split mandrel in a flex shaft.

3. Using glass brush, clean enamel under running water. Be certain to remove all flecks of material caused by grinding.

4. For last firing, add small balls on top. To add balls to each section, take a small amount of enamel that works with section color and mix with drop of Klyr-fire. Place some of mixture where silver balls are to be placed and then apply balls. Press balls down slightly.

5. Allow to dry completely, then fire. This firing will make piece glossy again. If there were no silver balls or other surface texture to add, then a hand-polish would be best.

Note: *Take care not to fire at too high a temperature or balls will sink too far into enamel.*

7. Clean up metal on front, back, and edges. See Photo #9.

8. Insert pin stem as shown in Diagram B. Bend pin stem at a 90° angle, leaving approximately ½" on end. While holding down short end to brooch back, bend pin stem straight up against hinge so it will spring upward and stay in catch. File to a point and adjust catch as needed.

Photo #7

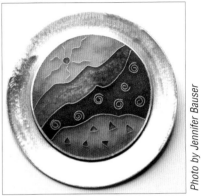

Photo #8

Photo #9

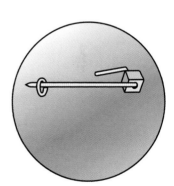

Diagram B

55

j.e.jasen

"Enameling and its connections to my life can be traced to my childhood, with my need to be near a hot oven and my desires to bake, . . . never, however, wanting to eat the things that I baked. The other major influence in my life could be my interest in patterns, textures, and colors in the world around me. I had a few interesting choices for careers—Enameling was the one which won out!"

What jasen truly loves about enameling is the fact that, like cake baking, enameling is very process-oriented and a slightly technical art form. One should follow certain recipes, but there is always room for adding her own special intuitiveness or something experimental.

In the past years, jasen has become increasingly aware of her environment, and become interested in ergonomics. Her experimentation in enameling is derived from her own curiosity for research with the materials, learning, and all that she has learned from the other artists working in compatible materials, such as the painter, the collage artist, the metalsmith, the ceramist, the glass artist, and the draftsman. Over the years, some of the compatible materials which can be observed in the patterns and textures of jasen's work include: luster materials, decals, ceramic pencils, china paints, underglaze paints, gold leaf, matte finishes, and gloss finishes.

Since 1979, as j.e.jasen, she has participated in international and national enamel exhibitions in galleries and museums, conferences and working symposia. Some of the more memorable ones have included:

** the "GOOD AS GOLD . . . Alternative Materials in American Jewelry exhibition sponsored by the Smithsonian Institute. This show (early 1980–85) traveled extensively throughout the North and South Americas;

** many international invitational enamel exhibitions and competitions in Canada, England, Finland, France, Germany, Japan, and the United States of America;

** and the International Enamel Symposium where she was invited to speak, participate, and work with other international enamelists from Austria, Germany, and Japan.

** Receiving a Grant from Craft Alliance of New York State, Inc.

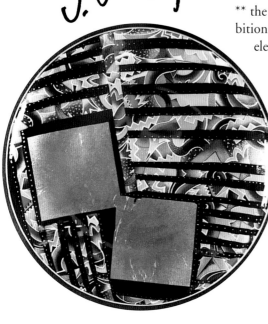

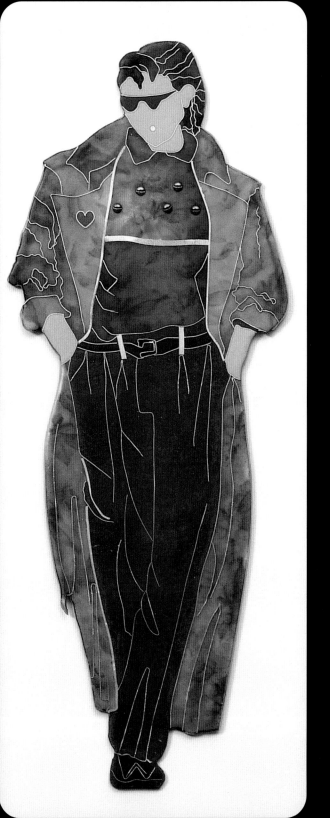

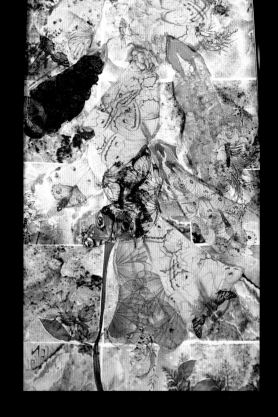

Decals

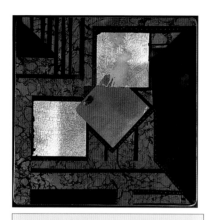

Techniques to Know

- Applying Enamels: Sifting, pg 31
- Counter Enamel, pg 35
- Preparing Metal for Enameling, pg 27
- Working with Foils, pg 36

Photo #1

Photo #2

Photo #3

TOOLS:
Cotton swabs
Craft knife
Dish of luke warm water
Firing tools
Rubber squeegee
Scissors
Straightedge for cutting
Tweezers

MATERIALS:
Copper piece to enamel
Enamels: 80 mesh, black opaque and artist's choice
Leafs: gold and palladium (optional)
Lusters: gold and palladium (optional)
Paper towels
Several sheets or patterns of decals

This project is about surface decorations, and combining enamels with low-fired ceramic materials and similarly low-fired glass-arts materials. These materials have been borrowed and adjusted from the other fired arts including: ceramic and glass decals, ceramic lusters and opalescence glazes, pencils, pens, crayons, and markers. China paints and over- and/or underglazes also fall into the general category of ceramic materials, because their original intent and their applications can sometimes be used on matted or glossed enameled surfaces.

The following instructions are being given for the enameling process. The instructions would differ tremendously if used in a ceramic or glass studio. The project discusses how to use decals, but there is also a short description of lusters and opalescence glazes, and ceramic pencils, pens, crayons, and markers.

Enameling:

1. Enamel both sides of copper shape to get a good base coat. See the Raku-fired Bowl project page 108. This project uses a black enameled base. Alternately, a pre-enameled piece can be used.

2. Sometimes decals are bought with protective paper on the face side. Remove protective paper. See Photo #1. Using scissors, cut or alter decal into desired shape and size.

Note: *Decals may be purchased at a ceramic or enamel supplier*

3. Soak in water with decal image facing up. See Photo #2. Remove when the decal lifts from the backing paper.

4. Using tweezers, lift decal from backing paper and place it on enameled surface. See Photo #3.

5. Gently smooth surface of decal to remove water from under decal and over enameled surface. A cotton swab, a rubber squeegee, or a matchbook cover may be used to flatten decal. See Photo #4. Dry excess water at decal edges with paper towels or, in small areas, use cotton swab.

6. Air dry. To speed drying, place piece on a grill on top of heated kiln.

Note: If air bubbles appear, place pin holes into the bubble and press decal down on surface. It should adhere.

7. Using craft knife, cut away excess decal that falls over edges. See Photo #5.

8. Place piece onto appropriate firing support and place in 1500°F furnace. After 30–120 seconds (depending on size of enamel piece and decal) combustion will take place. See Photo #6. It is not necessary to witness this part of process. Keep door closed during firing process.

Note: After top surface begins to melt, the bonding of decal to surface will take place. If fired too long after bonding, or if decal has been made with china paints, decal can possibly sink on the surface.

9. Apply other layers of enamels, lusters, and foils or leaf to design. Fire as needed. See Photo #7. All these elements can be added on one layer or multiple layers. In this piece, the leaf was applied with just a light spray of water, but holding agent can also be used.

Note: There are other ceramic materials that can be integrated in an enamel project, such as luster and opalescent glazes. These liquids can be applied with a brush or writing pens. These materials can be applied to any enameled and fired surface, but not directly on top of foils or decals. These glazes must be completely dry before firing and should be used on top of the final surface.

Other surface decoration materials include ceramic pencils, pens, crayons, and markers. The pens, crayons and markers can be applied to either a matte or gloss enameled surface, but ceramic pencils must be applied to a matte surface. Depending on the desired results, a piece may be considered finished after firing, or one last coat of clear flux enamel may be applied and fired on.

Photo #4

Photo #5

Photo #6

Finished piece

Photo #7

Charles Lewton-Brain

Master goldsmith Charles Lewton-Brain trained, studied, and worked in Germany, Canada, and the United States to learn the skills he uses. His work is concerned with process and with beauty as well as function. Much of it uses a printmaking approach to working metal, that is that the work is done in separate steps in groups building towards the finished piece. His work and writing on the results of his technical research have been published internationally. In 1991, with his partner Dee Fontans, he started a resource center and workshop school in downtown Calgary called the Lewton-Brain/Fontans Centre for Jewelry Studies, which brings in innovative exhibitions and educates metalsmiths.

A distinguished Fellow of the Society of North American Goldsmiths and a Fellow of the Gemological Association of Great Britain, Lewton-Brain has lectured and taught in England, the United States, Canada, and Australia. He regularly teaches workshops in the results of his research projects across North America. He developed fold forming—a series of techniques new to the metalsmithing field—which allow rapid development of three-dimensional surfaces and structures using simple equipment. The Rolex Awards for Enterprise chose a project of his on the further development of fold forming for inclusion in the *Rolex Awards for Enterprise 1991 Edition,* a book on innovative developments in science and invention in the world. In 1994, he founded Brain Press to publish *Cheap Thrills in the Tool Shop,* a book of inexpensive tool options and bench tricks for goldsmiths. Other books include *Small Scale Photography* and *Hinges and Hinge-Based Catches for Jewelers and Goldsmiths.*

Lewton-Brain served as a director on the board of the Alberta Crafts Council for five years and has now served five years as the National Crafts Representative on the board of the Canadian Conference of the Arts, where he has national reporting responsibilities and presents the views of Canada's national crafts organization, the Canadian Crafts Federation.

He has lived in Calgary since 1986 and is currently Jewelry/Metals Program Head at the Alberta College of Art and Design as well as writing articles, exhibiting, consulting, and creating pieces for sale. In 1996, he began a web site collaboration with Dr. Hanuman Aspler in Thailand. The Ganoksin web site is now the largest educational site in the world for jewelers, with over 30,000 individuals a month visiting and a 2,500 member e-mail list called Orchid.

Sterling silver fusion inlay under enamel

14k Gold fusion inlay under enamel

Basse Taille enamel pin

Fusion Inlay Under Enamel

TOOLS:

Brass brush
Firing tools
Finishing tools
Glass brush
Hemostat
Jewelry tools
Pickling tools
Rolling mill or hammer
Rubber gloves
Sifter: 200 mesh
Soldering tools

MATERIALS:

Binding wire, soft-iron. Only use wire that is easily dented with a hammer, 20ga or thinner.
Note: Other types of wire like brass or nickel silver will also work.
Borax-based brazing flux (nonfluoride is safest)
Car wax
Copper sheet: 1" sq., 8ga,
Counter enamel
Double-sided carpet tape, thin
Gold wire or strip: 2", 22k or 24k
Note: Can be done with 18k or 14k but results are not as nice
Manila file folder paper
Masking tape
Paper or label stock
Paper towel
Scalex
Transparent enamels: 80 mesh, browns, pinks, artist choice
White glue

Techniques to Know
• Applying Enamels: Sifting, pg 31
• Cleaning Between Firings, pg 28
• Cleaning enamels, pg 23
• Finishing, pg 37
• Metal Patterning for Basse Taille: Steps to roll-print onto metal, pg 26
• Preparing Metal for Enameling, pg 27

This project demonstrates the use of gold to create patterns and drawings under transparent enamel. This is a type of basse-taille technique and so works well if used with basse-taille texture methods in combination with the inlaid-gold patterns. In this project, a wire is pressed into a copper plate to create dents which are then filled with molten gold; this is called fusion inlay. The resulting pattern is then enameled over. There are other methods of using gold applications under enamel to make pictures not dealt with here (e.g.: fusion overlay, incised overlay, Keum-boo).

Metalwork:

1. Clean metal, then prepare a design, actual size, on paper to go onto copper square. If design is photocopied or laser-printed onto label stock it can be easily adhered onto copper square, or simply draw on the metal.

2. Prepare the copper square by pickling and rinsing, then brass-brushing and sanding the edges and corners smooth.

3. Using thin layer of white glue, glue line drawing onto copper square; allow to dry (with label stock, simply peel and stick). Cut and apply a piece of transparent double-sided carpet tape on top of drawing.

4. Using jewelry pliers, bend and cut binding wire to fit design and place onto drawing. See Photo #1.

Photo #1

5. Cover wire with masking tape and roll-print design into copper square. See Metal Patterning for Basse Taille, page 26. See Photo #2.

Note: *For illustrative purposes, masking tape has not been applied on top of wire in Photo #2.*

6. Remove all tape and wire from the copper. Using brass brush and soapy water, clean the copper and recesses of the dents. See Photo #3.

7. To melt gold into dents (fusion inlay), place copper square on light-type refractive firebrick. Flux dents and surface heavily.

Photo #2

8. Prepare gold wire by clamping into hemostat and fluxing it well.

9. Heat copper square with flame angled downward at corner closest to torch tip. See Photo #4. Do not remove flame from copper surface. Use large, intense flame. Use bottom half of flame (part furthest from torch). Heat until copper is glowing slightly and flux is well molten. Bring gold wire into flame and feed gold into the recesses (just like "stick" or "wire" soldering). Do not fill dents to overflowing, just near full. When gold has flowed in all recesses, stop heating. Move copper square as it cools down so flux does not stick to firebrick.

10. Pickle clean, rinse, scrub with brass brush, and dry. See Photo #5.

Photo #3

Photo #5

Photo #4

Photo #6

Photo #7

Photo #8

Photo #9

11. Clamp copper square in vise and file off any gold on top surface, leaving only that in recesses as an inlaid-gold pattern. See Photo #6. A belt sander may be used for this process. Check how much material is being filed by heating piece up a little and seeing how it discolors, or by painting surface with diluted liver of sulfur solution to see where gold lies and how much of pattern has been revealed. See Photo #7. This is the copper square after filing.

12. Again, use rolling mill to roll inlaid copper square, stretching it out in both directions, until it reaches approximately 18ga. See Photo #8.

13. Texture surface further with chasing tools, engraving tools, stamps, or use roll-printing techniques to produce decorative basse-taille effects to enrich existing inlaid design.

Enameling:

1. Prepare rear surface for enameling by glass-brushing with soapy water. Wear rubber gloves to keep from getting glass from brush in fingers. Dry.

2. Sift on counter enamel. Use Scalex on front, if desired.

3. Fire piece at 1500°F–1550°F for 1–3 minutes or until counter enamel is properly melted and smooth.

4. Pickle, rinse, and glass-brush front surface with soapy water.

5. Using 200 mesh sifter, sift an even layer of transparent color(s).

6. Fire piece at 1500°F–1550°F, but keep an eye on it. Underfire slightly—just when orange peel surface settles down. This provides a good contrast with the gold. See Photo #9.

Finishing:

1. Sand edges of plate and finish as desired. Burnish edges, then wax (to keep the metal from tarnishing). Mount plate in frame to finish. See Photo #10.

Photo #10

Enameling Photo Gallery

Examples of Fine Enamel Pieces from Professional Enamel Artists

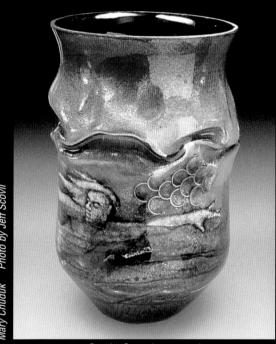

Mary Chuduk   Photo by Jeff Scovil

Stroke Camaieu vessel

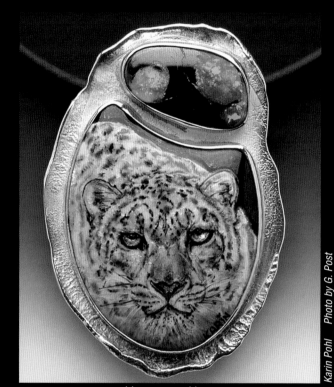

Karin Pohl   Photo by G. Post

Limoges necklace

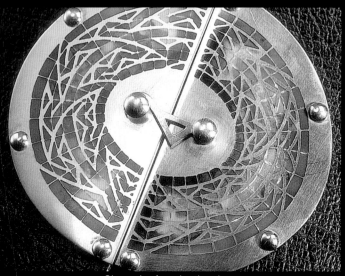

Joan MacKarell

Champlevé on leather box

Edith Koeppen   Photo by Tom Sommers

Cloisonné Pin

Joan MacKarell

Silver and enamel box

Coral Shaffer

Shaffer's enameling career began in the early 1970s. Using a borrowed kiln, she and two friends taught themselves how to enamel. Collectively known as "Enamelworks," they successfully sold their pieces at local art fairs. As they gained experience and became braver, they began making production champlevé and cloisonné pieces which they marketed nationwide and sold through stores and catalogs. After they had had five or so years of full-time experience, they contracted with Charles Scribner's Sons to write a book entitled *A Manual of Cloisonné and Champlevé Enameling*. It was published in 1981, but is now out of print. When lead-bearing enamels began to be hard to find, they decided to import some and another arm of their business, Enamelwork Supply Company, came into being.

In 1985, Shaffer was given a once-in-a-lifetime opportunity to study Japanese enameling techniques in Kyoto, Japan, at the Inaba Cloisonné Company. She studied there for 1½ years concentrating on several techniques that were new to her: Ginbari (embossed silver foil fired to a copper base and enameled), Musen (cloisonné with the wires removed before firing), and Shotai (cloisonné enamel with no metal foundation, achieved by etching away the metal base).

"I feel very strongly that information learned should be passed along. To that end, I have written several articles for *Glass On Metal*, and in 1990 I began teaching enameling workshops on a regular basis."

"In 1995, both of my partners left to pursue other careers; but I continue on, still mesmerized by the magic of glass on metal. Even after 25 years, I am NEVER bored—there are so many techniques to perfect and always new things to try. My days are now divided between enameling, teaching, and selling supplies. It's a great combo!"

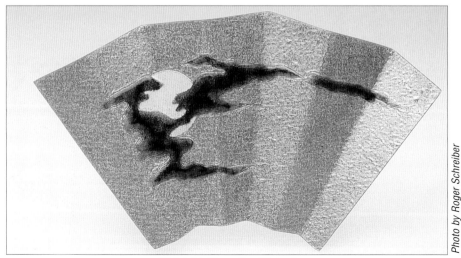

Photo by Roger Schreiber

Midnight Moon Fan Ginbari

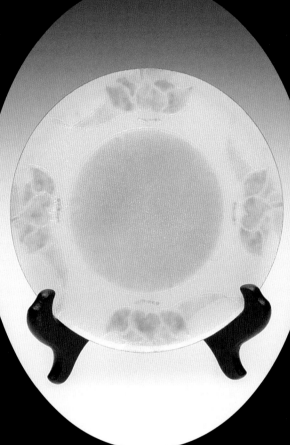

Photo by Roger Schreiber

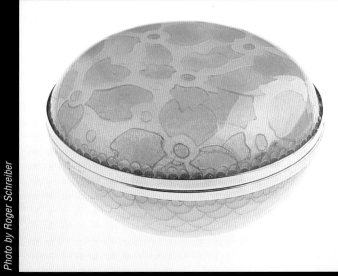

Photo by Roger Schreiber

Four Seasons Teapot Gold-plated cloisonné

Photo by Roger Schreiber

Small Covered Bowl Silver wire plique-à-jour

Ginbari Foil Embossing

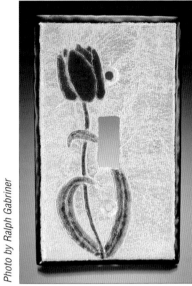

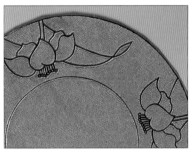

Photo #1

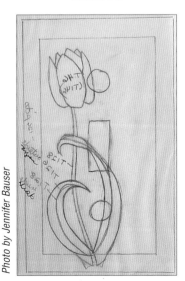

Photo #2

TOOLS:

Brayer
Cloisonné tools
Designing tools
Enameling tools
Felt scrap
Firing tools
Garbage can
Grinding stone
Heat lamp (optional)
Hot dog tongs
Metal snips
Pencil with new eraser
Sandpaper: 220-grit
Scissors
Sifter: 60 mesh
Tweezers

Techniques to Know

• Applying Enamels: Sifting, pg 31
• Applying Enamels: Wet-packing, pg 32
• Cleaning Between Firings, pg 28
• Preparing Metal for Enameling, pg 27
• Wirework, pg 29

MATERIALS:

Baby powder
Copper cleaner (optional)
Copper light switch plate or 18ga
  copper sheet cut to shape of the
  design
Copper wire: 24ga round
Cotton swabs
Enameling oil containing pine oil
Enamels: Thompson lead-free BC1070
  white (purchase dry), clear flux for
  silver, +200 mesh transparent colors,
  artist choice
Ginbari foil
Glue: Strong fast-drying
Holding agent (optional)
Mica
Pickle solution
Plastic or metal sheet, larger than
  design
Waxed paper

Ginbari foil embossing is a Japanese technique. Commercially, ginbari pieces are made using etched zinc plates for the embossing form and a printing press to emboss the foil. See Photo #1. Finished plate is shown on page 67.

Since zinc plates are difficult to obtain, the project and tools have been simplified so that ginbari can be easily done in a classroom or home. One of the benefits of the ginbari technique is the embossing form can be reused.

Metalwork:

1. To make the embossing form; first, draw a design. Flowing, organic lines are more effective in this technique than strict sharp-edged lines. See Photo #2.

2. Use cloisonné wire-bending techniques to bend 24ga round copper wire to match design lines.

3. Apply strong fast-drying glue to round wire and attach to a hard nonporous surface such as plastic or metal. If desired, transfer design to surface with carbon paper to make a guide for gluing wires. After wires are placed, cover form with waxed paper and weight it down for drying. This should make wires dry flat against surface.

4. After form is dry, shake on baby powder so that no sticky places remain. See Photo #3.

5. Clean copper switch plate thoroughly. To make a copper switch plate, cut out the desired copper shape, using 18ga sheet copper, and smooth edges. Drill holes for switch and screws. Dome piece. Clean as instructed above and prepare for enameling.

Enameling:

1. Mix white enamel to proper liquid consistency—approximately four tablespoons water with eight ounces of powdered enamel. Consistency is correct if enamel sheets without clumping when dipping a metal spoon into enamel and shaking off excess.

2. Holding piece by edges with tweezers or tongs, lower it into a container of liquid enamel until all areas on both front and back are covered with enamel.

3. In the center of a garbage can, twirl piece gently with a circular motion of the wrist to evenly distribute enamel. Dip again and repeat twirling motion.

4. Place piece on a trivet that touches only edges of piece. Dry. A heat lamp speeds up drying time. When piece is dry, it will look chalky, not shiny. Inspect for bumps in the enamel. These can be removed by gently rubbing with fingers or a brush. Place piece back on trivet.

5. Fire piece at 1500°F for approximately three minutes. Time will vary depending on many factors including size of piece and how quickly furnace reheats after opening door. Remove piece and cool. See Photo #4.

6. With sandpaper or a grinding stone under water, remove firescale from edges. This will need to be done every time piece is fired. Repeat Steps 2–6 to get a thicker coat of enamel.

7. Pierce ginbari foil to allow gasses to escape when firing. To properly do this, lay foil on piece of 220-grit sandpaper, lay a piece of felt over foil and sandpaper. Using brayer, roll over felt. Peel foil off sandpaper and look at it in front of a light source. Tiny holes of light should be evenly dispersed throughout the foil.

8. Place foil between pieces of mica to keep it from flying away in the kiln. Place entire mica/foil "sandwich" on firing rack. Anneal foil in the kiln at 1300°F for 2–3 minutes. Allow foil to air-cool.

9. Trace copper shape onto piece of paper. Place paper over foil and another piece of paper under it. Using scissors, cut out foil along traced lines. Remove papers.

10. Center foil over embossing form. Holding foil in place with one finger, cover foil with a piece of felt. Holding felt, foil, and embossing form steady, roll over felt with brayer. Carefully move hand so all areas are rolled. Still holding foil on embossing form, remove felt and look at foil. It should all be embossed. To further accentuate embossing, gently roll finger over foil at wire locations.

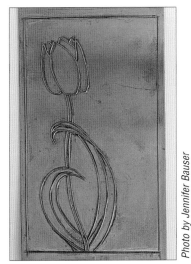

Photo #3

Photo by Jennifer Bauser

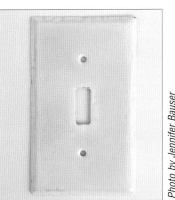

Photo #4

Photo by Jennifer Bauser

Photo #5

11. To attach embossed foil to the enameled piece, use a cotton swab to completely cover enameled base with enameling oil. Let dry approximately five minutes under a regular lamp or one minute under a heat lamp. Oil should become dry but tacky. Carefully center foil above enameled base and lower it down into position. Once it has touched enameling oil, it will not move again.

12. Using a pencil eraser filed down to a point, press all the flat foil areas down against the base, using a tapping motion. Follow along both sides of every design line with eraser tool, making certain foil is sticking to base. Be very careful not to touch raised design lines or they will be crushed. Dry for a few minutes.

13. Place piece on trivet and fire at 1500°F for approximately five minutes. Enamel should get molten enough to be drawn up into raised areas of design. When piece is cooled, press down on one of the lines with a fingernail. It should be firm and not give with pressure. If lines are not firm, refire for a longer time. See Photo #5.

14. From this point on, ginbari is treated like any other wet-packing technique. Clean enamels and wet-pack them wherever appropriate. Mix holding agent with enamel/water solution to keep enamel in its allotted space. Fill in all desired colors in this manner.

Note: *If using colors that change over silver, be certain to apply a coat of flux before applying colored enamels.*

Note: *The design will be more prominent if the raised lines are not covered with colored enamel.*

15. Place piece on trivet and allow enamels to dry. Fire at 1400°F for three minutes. Cool.

Finishing:

1. When coloring of the piece is finished, paint enameling oil over entire front of piece and wipe it off with a tissue. Using 60 mesh sifter, sift clean dry clear-enamel flux for silver over enameling oil. Turn piece upside down over a piece of paper. All but a thin coat of dry enamel will fall off. This last layer of flux will inhibit the foil from tarnishing. Fire as before.

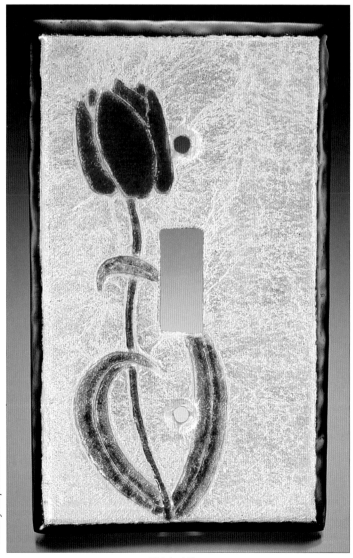

Finished Piece

Enameling Photo Gallery

Examples of Fine Enamel Pieces from Professional Enamel Artists

Geraldine M. Berg

Marianne Hunter   Photo by G. Post

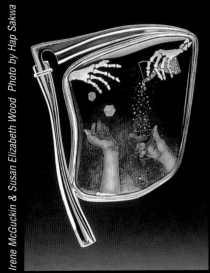

Irene McGuckin & Susan Elizabeth Wood   Photo by Hap Sakwa

Hand me The Wine and The Dice

Isabella Corwin

Fossil Fish

D.X. Ross

"One of my earliest memories is of drawing trees in winter, branch by branch, exploring the multiplicity developing from a single trunk to hundreds of twigs. I have been drawing ever since. When I took a survey course on enameling in college, I bonded with the technique of grisaille. Unlike works on paper that need to be framed or bound or carefully stored, my enamels can be made into jewelry!"

Since grisaille is so refined, Ross is able to draw anything she imagines in a small area. She began using the wonders of nature and mythological creatures as her subjects. Egrets, for example, symbolize elegance plus the magical skill of flying. And Mercats are what Mermaids have for pets. How do you feel if you are wearing the image of beauty and power of flight? How do you feel if your cat-pet has the tail of a fish? You may feel that you are elegant and omniscient and specially gifted.

Ross's current series is jewels for gardeners—flower and leaf forms found in the landscape of her mind. By pinning her fabricated flora to lapels, she hopes to evoke how creativity celebrates life.

DX

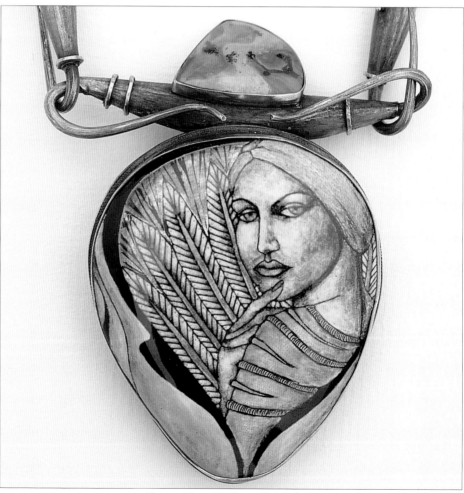

Ocean Thoughts Domed pendant

Daneen Sterling crystal

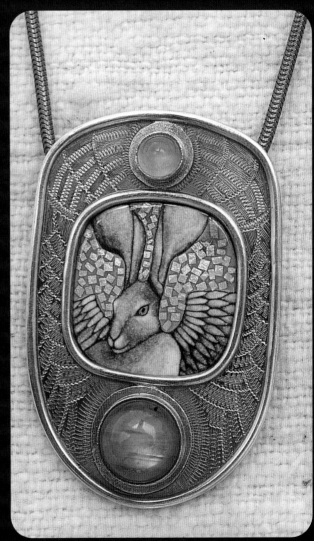

Magic Rabbit Enamel

Magic Cat Enamel, silver, moonstone

Grisaille

Techniques to Know
- Applying Enamels: Sifting, pg 31
- Cleaning Between Firings, pg 28
- Preparing Metal for Enameling, pg 27
- Sgraffito-See Sgraffito Plate project, pg 118

Photo #1

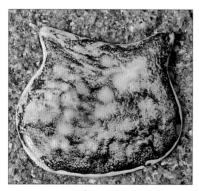

Photo #2

TOOLS:
Enameling tools
Finishing tools
Firing tools
Grinding stone, artist choice
Paintbrush: 00, #6 sable
Pickling tools
Scribe
Sifter: 80 mesh

MATERIALS:
Enamels: Black, +200 mesh hard-fusing, if possible; white, -325 mesh; pink, -325 mesh; green, -325 mesh
Klyr-fire®
Liquid detergent

In grisaille, images are formed by many thin washes of white enamel over a base coat of black enamel. The small particle size of the white enamel allows fine detail and shading, the many applications and firings produce depth and magic. Designs that work best include shading that describes volume, as in drawing an egg with light, shadow, and the gradation of grays between them; and designs that have areas of pattern such as fishscales. A fish that is drawn with scales emphasizing its contour, and then shaded accordingly, makes a stunning statement in grisaille. However, a flat checkerboard of highly contrasted black and white is not a candidate for the grisaille technique as it requires sharp edges and no softness between the black and white.

Enameling:

1. After the copper is formed and edges filed, prepare metal for enameling by annealing. Cool. Soak in pickle of salted vinegar. Wash with toothbrush and liquid detergent.

Note: After each firing, stone edges under water and wash with toothbrush and liquid detergent.

Note: When a wet enamel medium is needed, use a 1:1 Klyr-fire/water solution.

Note: Wear face shield and protective glasses when looking into furnace.

Note: Test the firing temperature of the furnace; heat should be high enough for the piece to fuse in 1–2 minutes (approximately 1500°F).

2. Sift black enamel for counter enamel. Fire. Cool.

3. Stone edges under water, soak in pickle, then wash as in Step 1 above.

4. Sift black enamel on front. Fire. Cool and clean as in Step 3.

5. Sift second coat of black on front side. Fire, cool and clean edges. See Photo #1.

6. Paint a thin wash of white over black shiny surface. See Photo #2. To do this, make mixture of enamel powder in 1:1 Klyr-fire/water solution. Do not mix large quantity because extra dry powder and liquid on the side is needed to vary intensity of wash, as in watercoloring. For deeper color, add more white enamel, for paler or blended areas, use more Klyr-fire/water solution to thin pigment. Alternately, brush piece with Klyr-fire, then sift thin layer of white enamel through 300 mesh

screen. The correct amount of white enamel will come with experience; too much limits the range of grays possible, too little will make the sgraffito difficult to find after firing. Dry.

7. Using scribe, scratch lines through dried white enamel before firing. See Photo #3. Black enamel is not applied again. Fire when all black lines and black areas are clear of white. This firing is critical. Heat until enamel goes from white-and-chalky looking to wet-and-grainy looking. **Note:** *This happens quickly so be attentive.* Remove immediately before enamel glosses. Cool. See Photo #4.

8. Paint white enamel in all places that deep, pure black is not desired. See Photo #5. This is the easiest time to find background areas that are dark but not pure black. Visually, these areas will recede into background if they are worked early in the process and left without addition in later applications (this is the strategy for portraying illusion of depth). Otherwise these areas may appear too strong, making finished design look flat. Use heavier amounts of white in lighter areas, thinner in shadows. Use shiny contrast of black lines and areas to graininess of underfired white layer to find drawing by angling piece in the light. It is impossible to keep wet brush strokes of enamel completely inside lines. Excess must be cleared away with scribe. See Photo #6. Scribe is able to smoothly slide along black lines and margins, but it snags in grainy white surface. Fire carefully, until dry enamel becomes wet-looking, but remove from heat before it glosses to take advantage of textural contrast for next application. Cool. See Photo #7.

9. Continue to add layers of white enamel wash—thicker where highlights and brighter elements are, thinner and eventually not at all where background and shadows are. Remember to clear all excess enamel from lines and edges. After first couple of firings, the contrast in value is great enough to see, so underfiring is no longer necessary. See Photo #8. Notice green eyes and pink nose. See Photo #9. Apply opaque colored fines (-325) near the end of the process. Approaching completion only highlights need white.

Many careful applications and firings are required to build up full range of grays from black to white. See Photo #10. This may take 10–15 layers. Larger pieces of at least 4" square will take 20 or more layers. After last firing, no stoning is necessary. Set or mount piece as desired.

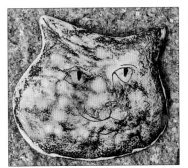

Photo #3

Photo #7

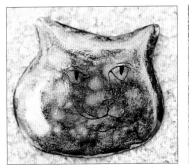

Photo #4

Photo #8

Photo #5

Photo #9

Photo #6

Photo #10

Ora Kuller

"Enameling for me was a coincidence, and it was not love at first sight. I was studying at an art college in Israel at the time and took the enamel class as it fitted nicely into my schedule. I did not like the enameling class too much. I had the feeling that working with enamel was a tedious, dusty, and limiting kind of art form."

"When I had to attend the last enamel class of the year, I drove to school very reluctantly. It was a hot summer's day and nobody else bothered to come to class. Sitting at the bench, cluttered with mosaic stones, pieces of glass, and dusty jars of enamel, I started to transfer an image from my mind to some pieces of copper. The room was quiet, the kiln was mine, and the magic was happening. From the dark surface of the metal, the dreamy image of a white heron slowly peered at me—white throat, white head, red beak, shiny eyes, green and blue water. I gazed, transfixed, at my creation. Is that how it felt to create the world?"

"At that moment I knew what I wanted to do with my life."

After creating the white heron, Kuller has discovered countless ways to express her thoughts and dreams in enamel. She loves the softness and the subtle nuances of watercolors and was thrilled to find similar possibilities within the enamel world.

Watching the delicate lines and features of her daughters and her mother appear on the perfect enamel surface is like magic. Often when she walks through museums or leafs through art books, or just reads books, some expression—a movement or a certain color in an artwork—catches her imagination. She tries to portray one of her daughters or just create an enamel artwork that will be reminiscent of these feelings. Other times, some elusive spirit in Kuller's drawings will have so much magic in it that she just has to turn it into enamel. She will follow this magical feeling with research and more drawings until her inner vision and the outcome on the enamel surface satisfies her.

One other passion Kuller has is to design miniature enamels, set them in silver and copper, and employ them not only as the jewelry pieces that her images wear, but also as the connecting elements that hold the entire work together.

Enameling is an exacting art. The combination of metal and glass, both rigid materials, forces Kuller to make early choices and work within set boundaries. The trick is to know how to bend these two demanding and unforgiving substances into the composition of shape, line, color, and mood that her soul desires.

"I love knowing that enamel artists in ancient times worked in very similar ways to mine. Those ancient artists, like myself, knew they had to invest long hours of meticulous work in order to have the glorious chance of conveying their ideas through this shiny, iridescent material that never ceases to amaze."

My Cup of Tea

Detail *From My Cup of Tea* Enamel, fine silver, moonstones

Limoges—Painting with Enamels

Techniques to Know
- Applying Enamels: Sifting, pg 31
- Cleaning Between Firings, pg 28
- Counter Enamel, pg 35
- Finishing, pg 37
- Liquid Enamel, pg 35
- Preparing Metal for Enameling, pg 27
- Test-firing Colors, pg 24
- Grade-sifting and Particle Sizes, pg 20
- Sgraffito-See Sgraffito Plate project, pg 118

TOOLS:
Alundum stones: 150- and 200-grit
Ballpoint pen
Candle and matches
Clay shapers or fine brushes
Drill and bits
Eyedropper
Firing tools
Jeweler's saw and blades: 2/0, 4/0
Metal file
Paintbrushes: 20/0, 10/0, 3/0, 1
Photocopier
Scribe
Sifters: 60, 100, and 200 mesh
White china plate

MATERIALS:
Copper sheets: 6" sq., 18ga (2)
Cotton balls
Distilled Water
Double-sided tape
Enamels (see chart below)
Fine-silver sheet: 6" sq., 20ga
Klyr-fire®
Masking tape
Metal-cleaning polish
Powdered kitchen cleanser
Scalex
Silicone caulking

Thompson lead-free enamels:		Overglaze Painting Colors:
Name	**Type**	
533 white (counter enamel)	T	906E green, 907E red, 908E yellow,
1010 undercoat white	O	909E orange, 910E brown, 911E blue,
1440 delft blue green	O	912E black, 913E mixing white*,
1685 cobalt blue	O	914E white liner, 1705P petal pink,
1912 nude gray	O	1708P pastel pink, 1715P clover pink,
2010 soft fusing clear	T	PF-1 painting flux*
2020 clear for silver	T	
2115 mars brown	T	* these two colors are VERY
2615 periwinkle blue	T	important to the project

Underglaze black (oil base)

Ceramic Pigments:
OC-16 blue, OC-18 blue, OC-195 blue, OC-32 yellow, OC-70 red, OC-71 orange, OC-82 yellow brown, OC-83 brown, OC-85 red brown, OC-95 red purple, OC-169 tan yellow, C-170 blue green, OC-191 green.

This project is not for the faint of heart—painting skills are needed for shading and building up the design. The technique, however, is fascinating and rewarding as the finished piece can be similar to a fine painting.

The project is done with overlapping pieces, then assembled at the end. This gives dimension to the work and allows the use of a smaller furnace. Mounting considerations are given at the end of the project. Different particle sizes are used in this type of work. Grade-sift all enamels, using a stack of sifters with 100, 150, 200, and 325 mesh sizes.

To simplify this project, ceramic pigment and enamel watercolors have been used to paint the design. However, the hair is painted and sgraffitoed through an underglaze. Foils and other decorative additives could also be used if desired.

Design:

1. Use a fascinating photograph for a pattern. Painting seriously in enamel is a long process, so select an interesting subject. See Photo #1.

2. Sketch design in detail. Make certain the whole design is finished before cutting or enameling are started. See Photo #2.

3. Photocopy design. Divide design into multiple pieces, and decide which will be cut in copper and which in fine silver (see Note below). On copy, divide design into segments, showing with dotted lines how they will overlap. Decide where mounting holes will be located because they need to be drilled before enameling (see Mounting Considerations on page 84). Label each piece with a letter. Make several photocopies of layout. This project has 13 segments, labeled A–M. Fine silver is used for scarf segments and copper for all other segments. See Diagram A.

Note: When deciding what metal to use for each segment, think of character, color, and shape. If character of segment is elegant, aloof, and very modern, or if cool colors are used, or if shape is intricate, use silver. Silver is cooler, more elegant, and it does not generate firescale. No firescale means no cleaning and filing of edges. It means more speed in work and less damage to delicate shapes of metal, as no file is needed to clean edges. On the other hand, if color scheme is warm with a straightforward shape and the subject, character, and atmosphere are reminiscent of antiquity or renaissance, copper works very well.

Metalwork:

1. Cut paper design into planned pieces. Cut at least two copies. One set is used for metalwork, another for stencils or for use in blocking out part of the work.

2. Cover front sides of copper and silver sheets with double-sided tape. Place segments, as appropriate, on exposed double-sided tape covering copper and silver sheets.

3. Using jeweler's saw with 2/0 blade, cut copper sheet according to design. Using 4/0 blade, cut the more intricate pieces from silver sheet.

4. Lay out all segments according to design and see if it they fit properly. If not, recut poor-fitting piece.

5. Drill connection holes into metal.

Photo #1

Photo #2

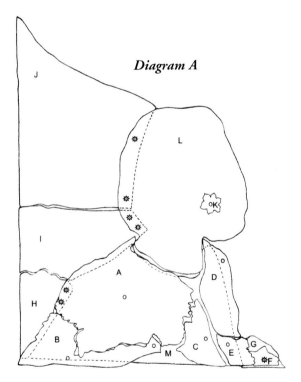

Diagram A

Note: In Diagram A, notice that pieces do not fit exactly. This is part of the challenge of creating a design with metal and glass. The saw blade may move, the heat of the furnace changes the size of metal a little; a certain segment may warp during firing. Such occurrences will happen and must be used as part of the project. Do not always place pieces of metal side by side but rather, let them overlap a bit.

79

Enameling:

Each segment of project will take similar steps to complete. Here is a general overview:

aa. Grade-sift appropriate colors. Some will be needed in +100 and -150/+200 mesh sizes. Straight 80 mesh is also needed.

bb. Clean metal pieces in preparation for enameling.

cc. Counter enamel Layer 1. Apply one layer of liquid counter enamel and allow to dry.

dd. Prepare background on front side for painting by sifting a smooth surface of a background color of granular enamel. This is done using a 60, 100, and 200 mesh sifter method for first firing: first sift through a 60 mesh sifter, then a second layer very lightly sifted, through 100 mesh sifter and last layer very lightly sifted, through a 200 mesh sifter over entire surface and especially over edges. If appropriate, after carefully lifting stencil that masked out an area, clean the space of any specks of enamel using clay shapers or fine brushes. Fire, cool.

Note: Holding agent is not used unless area to be covered is domed.

ee. Additional enameling may be done on front. Fire at 1500°F until glossy.

ff. Counter-enamel Layer 2. Sift, using 2020 clear for silver. Fire at 1500°F until glossy.

gg. Stone flat the painting background on front. Use an alundum stone, 150-grit for high places and 200-grit for everything else. Clean well under water, then dry. Background area is sifted again, lightly, using only 100 and then 200 mesh sifters. Fire at 1500°F until glossy.

Note: The tiny particles of enamel fill every space and the result should be a smooth, even and clean surface. If not, repeat this step until a smooth, clean surface is obtained.

h. Front is painted with enamel watercolors. Fire as needed.

Note: All layers of painting are underfired, until end, as watercolors can only stand limited heat and firings.

Note: On copper pieces, between firings, file edges to remove firescale and, if present, remove firescale inhibitor. Remember the second set of cut segments from pattern will be used as stencils.

Head: segment L

1. Clean and apply counter enamel to Layer 1 as in Enameling: Steps bb and cc.

2. While blocking hair area with appropriate paper stencil, sift 1010 undercoat white over face and neck areas, using 60, 100, 200 mesh sifter method as in Enameling: step dd above. Without touching sifted white enamel, paint hair area with Scalex. Let dry, then fire for 1½ minutes at 1500°F—or until white enamel is glossy. Cool.

3. Clean firescale off edges and remove Scalex.

How firing affects watercolors

Watercolors before firing

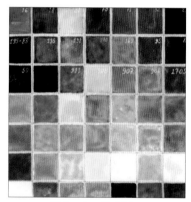

Watercolors first firing

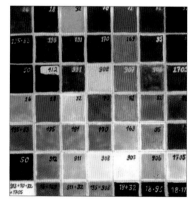

Watercolors second firing

4. Apply a very thin layer of underglaze black to hair area. Let dry. Since the underglaze base is oil, either allow a long period of drying time, or place piece on top of furnace. When underglaze is dry, sgraffito pattern of hair with scribe. See Photo #3. Look at design of hair—brown area is where the underglaze black was sgraffitoed.

5. Using powdered cleanser, clean paintbrush—but never use this paintbrush for applying enamels.

6. Using face/neck stencil to block all but hair, sift +100 mesh 2115 mars brown with 60 mesh sifter over hair area. Remove stencil and clean off all enamel that may have fallen into face/neck area. Fire at 1500°F until shiny and transparent.

Note: Always fire underglazes with transparent enamel over them.

7. Counter-enamel Layer 2. See Enameling: Step ff on page 80.

8. Stone white enamel. Using hair stencil, sift 1010 undercoat white using 100, 200 mesh sifters method as in Enameling: Step gg. See Photo #3.

9. Prepare colors for painting.

Photo #3

To prepare watercolor-type paints:

Overglazes are a way to paint with watercolors over enamel and gain all the magic that enamel offers. Use three types of paints: (1) The fines of the enamels (-325 or 400 mesh); (2) overglaze painting colors; (3) ceramic pigments. All are in powder form. #1 and #2 are enamels, but #3 are not enamels and are concentrated. Ceramic pigments must be mixed to compensate as follows:

For each color, mix 1:4 ceramic pigment to either painting flux or mixing white. Mix only a small amount at a time. Both painting flux and mixing white come as powders. They are both enamel and will make the pigments shine on firing. Painting flux will retain the intensity of the color, mixing white will make it softer.

Liquefy these components into watercolor-type solutions. Add two drops of Klyr-fire and two or three drops of distilled water to each color, then stir for a few moments until no lumps are left (lumps do not fire well and will not be the same color as the rest). The mixture should be thin enough to allow the use of a brush as in watercolors.

If these colors have never been used before, make color charts of the mixtures with painting flux and those with mixing white. Label with their numbers.

10. When transferring drawing onto enamel surface, do not use commercially available materials because they may harm enameled surface. Instead use candle soot. Hold a white china plate above a candle at an angle that will allow the flame to cover the plate with black soot. See Photo #4. Wipe soot onto a cotton ball and rub it on the back of one of the copies of the design. In this project, only do this for segment L.

Photo #4

81

11. Tape blackened side of paper on top of white surface of the enamel to work area so it will not move. Using a ballpoint pen, trace carefully over the lines. The transfer is very faint and only hints at important features.

12. Using 20/0 or 10/0 paintbrushes, paint over faint lines with 2:3 mixture of OC-195 blue and OC-83 brown (see Note below). Look at design to correct lines if needed. Take advantage of the enameled surface. If lines are incorrect, simply wipe them off. Refine drawing by cleaning lines with moist brush or sharp scribe.

Note: Do not use black enamel. Mix dark color by combining OC-195 blue and OC-83 brown in 2:3 ratio. This is a color that stays dark and crisp for many firings. It can also be used for shadows over the face.

13. When the outline is perfect, underfire piece in a 1450°F–1500°F furnace for 80 seconds or less, depending on size of piece. The lines will get darker, and when cooled, lines can be felt with the fingers.

Photo #5

14. Start adding shadows. Add deep shadows with blue/brown mixture. By adding blue/brown mixture to 913E mixing white, shadows can be softened in areas where needed. Nearly all enamel watercolors fire to a darker color than expected. It takes many layers, paintings, and firings to make shadows over the face deep enough and soft enough. See Photo #5.

15. Soften shadows with a skin-tone mixture: 913E mixing white with tiny amounts of OC-70 red, OC-32 yellow, and 1705P petal pink. White should be the dominant amount in this mixture. Add color to eyes and lips.

16. Fire only until colors darken inside furnace and take out before they get glossy. Paint as much as possible between firings. Fire only when absolutely necessary.

17. When almost finished, sift entire face with skin-tone mixture and fire at 1500°F a little longer, but not to glossy yet.

18. Add all the little details such as eyelashes, etc. Fire at 1500°F until glossy (approximately 1½ minutes). See Photo #6.

Photo #6

Note: Never judge firing time solely by the clock. Use the clock as a guideline, but refer to the color of the furnace and the work inside.

Shirt: segment M

1. Clean and counter-enamel Layer 1 as in Enameling: Steps bb and cc on page 80.

2. Sift 1685 cobalt blue on front first, using a 60 mesh sifter and then again, lightly, with thin layer, using a 200 mesh sifter. Fire until smooth and shiny.

3. Counter-enamel Layer 2 as in Enameling: Step ff. Fire.

4. Stone piece, if necessary, and sift over entire surface with 2615 periwinkle blue using +100 and -150/+200 mesh enamel. Fire for two minutes at 1500°F or until glossy.

Scarf: segments A, B, C, D, and E

1. Clean and counter enamel Layer 1 all pieces as in Enameling: Steps bb and cc on page 80.

2. Using paper stencils, block off leaf areas, sift 1010 undercoat white enamel on all parts that are meant to be flower petals, using 60, 100, 200 mesh sifter method as in Enameling: Step dd.

3. Using paper stencils, block off flower areas, sift 2010 soft-fusing clear enamel on leaf parts, using 60, 100, 200 mesh sifter method.

Note: 2010 soft-fusing clear enamel turns transparent yellow gold over fine silver. See Photo #5 for differences between the petals and the leaves. Fire until glossy.

4. Counter-enamel Layer 2 as in Enameling: Step ff. Fire.

5. Sift a delicate soft veil of -200/+325 2020 clear for silver through the 200 mesh sifter over the piece for every firing from now on.

Note: The underfiring technique, learned from Rebecca Laskin, should be applied at least twice to be beautiful and not too many times (maybe eight maximum). If work is far from completion, consider firing it without the thin clear coats until the last two firings. All the fines of enamel when sifted through a 200 mesh sifter, help in creating this wonderful texture in the underfiring technique.

Photo #7

6. Using blue/brown mixture, paint lines and deepest shadows of petals. Using 914E white liner, paint the veins in leaves. Underfire at 1500°F. See Photo #6 on page 82.

7. Enhance shadows and add more lines if needed. Soften interior veins of petal, by painting 913E mixing white over them. Leave lines that define borders, dark.

8. Using stencil over leaves, and using a 200 mesh sifter over flowers, sift thin layer of 2020 clear for silver. Fire pieces individually at 1500°F until the image emerges from beneath layer of sifted enamel. Remove from furnace. The underfired surface, when cooled, should look and feel like wet sand. This effect usually takes between 45–70 seconds in furnace.

9. Paint subtle shadows, lights, colors, and lines over underfired surface. Paint at least six layers of color with 2020 clear for silver sifted between them. See Photo #7. Watch the different stages in painting of scarf. Wet-sand texture is used for final—these segments do not get fired to glossy.

Remaining segments: F, G, H, I, J, and K

Note: F and G are not shown in Photos 3–7

1. Clean and counter-enamel Layer 1 each piece as in Enameling: Steps bb and cc on page 80.

2. On segments F, G, and H, sift 1912 nude gray; on segment I sift 1440 delft blue green; on segment J sift 1010 undercoat white. Use the 60, 100, 200 mesh sifter method as in Enameling: Step dd. Fire to gloss.

3. Counter-enamel Layer 2. Fire

4. Stone each piece as in Enameling: Step gg, then sift, using appropriate colors on each piece, using 100 and 200 mesh sifter method. Fire. Repeat as necessary.

5. For segments F, G, H, painting of rocks and pebbles on the seashore are done in much the same way as the scarf and has the same texture.

6. Segment I: the sea. Over background, paint deep shadows of the water, the highlights of the foam over the waves, the colors that play over the sea, the mountains and the forests on the narrow strip of land in the horizon. Use OC-195, 16, 18, 83, 32, 71, 914E, and 911E. When finished painting, sift +100 mesh of 2615 periwinkle blue over piece, but not on mountains, and fire. Add more colors and fire again. Do at least three layers.

Photo #8

7. When painting of the sea is satisfactory, sift 2020 clear for silver through a 200 mesh sifter, only on the mountains and forest strip, then fire a short time for sand-like texture over that particular area.

8. Segment J: the sky. Begin as with the face, only do not draw with lines over white enamel, but rather in masses of colors. Paint design of clouds in sky adding blues, yellows, whites, and grays. See Photo #8. Photo was taken before watercolors were fired in areas of sky and sea.

9. To soften the sky, use a 200 mesh sifter and apply a thin coat of 1010 undercoat white enamel, then fire a short time for sand-like texture. Repeat if needed.

10. Segment K is the earring that is used as a decorative and connecting element. Shape piece of 22ga fine-silver sheet into a flower. Drill hole in middle of flower. Sift 2010 soft-fusing clear on back of flower, fire and cool. Repeat this step for front.

Mounting Considerations:

1. When planning the design, determine exact location of holes for connecting elements. Also, decide to what the enamel will be attached. The project piece is connected to a panel of plywood covered with velvet fabric.

2. Connecting elements can be hidden from sight, disguised, or made into an important element of the design. For example, one connector is in the center of the earring, which disguises the element.

3. The simplest connecting element is made from fine-silver wire. With a torch, ball the end of the wire. This can now become a connecting element (a nail). It can be oxidized, enameled, or used as is.

4. If connecting elements are hidden, simply screw through holes in the enamel onto the wood behind. Drill holes in copper or silver sheet large enough to accommodate at least the size of miniature screws.

5. Make camouflaged elements. Solder a tiny piece of metal, copper or silver, onto a metal wire with hard solder. Enamel the head in the spirit of the work. Thread wire through hole in enameled metal and through the wood and connect it to a screw on the other side of the work.

Photo #9

6. To raise a certain portion of the work, use balsa wood, which is easy to shape and cut. Make it a little smaller than the segment. In this project, thin balsa is used under the head. The shirt is attached to a thicker piece of balsa wood so it is placed above the face.

7. Holes in the metal should correspond to holes that are drilled in the wood.

8. The wood panel may show in the design or not. If part of the wood is seen, choose quality wood. If covering the wood with fabric that fits the colors and texture of the work, use plywood. See Photo #9. Plywood is a good choice when design covers the entire panel of wood; cut panel smaller than the design if it is not to be framed. If work is to be framed, allow plywood to extend past design. The frame will hold onto the margin of wood around the design. See Photo #10.

9. Use silicone caulking in addition to the connecting element to fasten pieces to wood backing.

Photo #10

Judy Stone

Stone has been working in her medium since 1972 and is basically self-taught. She enamels copper objects which she fabricates by sawing sheet copper and either raising it or pressing it in a hydraulic press. Stone's studio is surrounded by a somewhat carefully cultivated garden from which she draws inspiration. The imagery in her work, while mostly abstract, has reference to forms lurking in her subconscious. She does not try to interpret these forms or their interactions with each other.

Over the years Stone has developed her own way of working based on medieval enameling techniques, the availability of enameling supplies, and the contemporary work of the late Fred Ball. She works in layers, pushing the enamel particles into place and then firing at between 1450°F and 1500°F. Most of her work is fired between six and fifteen times. Stone also incorporates glass balls from Japan and other compatible glasses into her pieces and, in some of her work, she fuses precious metal foils into the final layers. She tries to use the interaction of the metal and the enamel through transparency and translucency to achieve a three-dimensional color.

"The objects I make are both functional (light switch covers) and nonfunctional (vessel forms and wall pieces). My enameled light switch covers have been called 'site specific functional art.' I take pride both in the beauty and the functionality of these pieces."

© stone '01

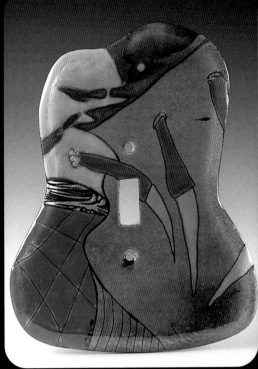

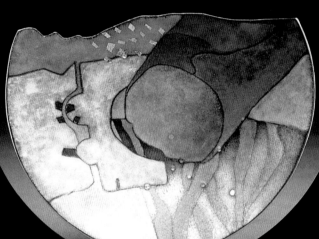

Liquid Enamel and Glass Ball Additives

MATERIALS:

Copper plates: 5" x 5¼", 18ga, with evenly beveled edges (3)

Enamels: 80 mesh, medium fusing white or flux for counter enamel; 80 or -150 mesh leadfree transparents; 80 or -150 mesh flux; -325 mesh lead-free opaque

Glass balls: red and light turquoise

Liquid counter enamel: either white or flux

Liquid enamels: white, flux

TOOLS:

Air compressor

Airbrush (double action preferred) with wide tip and needle for spraying glaze-type material

Bamboo skewer

File: halfround medium/fine

Firing tools

Firm, heat-resistant surface

Heat gun

Paintbrush: inexpensive

Paintbrushes: 00 to 000, synthetic sable

Pickling tools

Point rack to fit piece to be fired (points should be covered with kiln wash)

Respirator mask

Sifters: 80 mesh, 150 mesh, 200 mesh; 1"–2" diameter

Spray booth (can be as simple as a cardboard box with a hole cut in the rear and an exhaust fan mounted to draw out overspray through furnace filter)

Steel press plate

Toothbrush

Tweezers, curved

Watercolor palette with enough compartments for enamels used in project

Photo by G. Post

Techniques to Know
- Applying Enamels: Sifting, pg 31
- Cleaning Between Firings, pg 28
- Counter Enamel, pg 35
- Grade-sifting and Particle Sizes, pg 20
- Preparing Metal for Enameling, pg 27
- Sgraffito-See Sgraffito Plate project, pg 118

This project will show how to use liquid enamels as base coats, techniques for using different particle sizes, and how to adhere glass balls to an enameled piece. Particles sizes seem to cause some enamelists confusion. This project will use -150 mesh transparent sifting to get a watercolor-type wash effect and to achieve a maximum pull-through effect on the layer under transparents. Additionally, because these pieces will be fired several times, clear leadfree transparents will eventually develop. Use -325 mesh leadfree opaques because they act almost like opalescents when applied in this manner.

Metalwork:

1. Degrease metal, preparing it for enameling.

Enameling:

1. Place tiles face up in spray booth. With double-action airbrush, spray liquid white enamel evenly over front of tiles. Liquid enamel will first need to be stirred until smooth and pressure regulator may have to be adjusted to account for heaviness of liquid slip. Airbrush needle can be backed off to allow for greater flow.

Note: If airbrush, compressor, or spray booth is not available, an even coating of liquid enamel may be applied with soft watercolor brush. Use paintbrush to pull a little water over top of brush strokes to smooth enamel.

2. Using standard heat gun turned to high, dry liquid enamel. This drying process is sometimes called bisquing because it makes enamel surface harder than conventional air-drying. When pieces are cool, gently remove excess enamel on sides and place pieces on clean paper ready for sgraffito.

3. With a sharpened bamboo skewer, sgraffito lines and negative space of image on each tile. Gently knock off excess enamel powder on surface. Broaden lines and increase negative space by brushing with soft synthetic sable paintbrush.

4. To blur sharp line of hats that are in negative relief in top half of Tile 3 below, gently wet surface with water from an atomizer or airbrush (with the compressor set on low pressure and airbrush needle pushed all the way into tip) until edges begin to blur. Dry tile with heat gun as described in Step 2.

5. Place tiles on clean piece of paper and apply liquid flux enamel with paintbrush in areas meant to represent water. Keep liquid flux from touching dried and sgraffitoed white. Dry and sgraffito as in Step 3. See Photo #1.

6. Apply liquid counter enamel on back of each piece with paintbrush and lightly sift over wet counter enamel with the same color 80 mesh enamel. Do not use heat gun on backs of pieces because it will dislodge loose sifted particles.

7. After backs of pieces have dried, wipe off any counter enamel which is on edges of tiles.

8. Place tiles, one at a time, sgraffitoed side down on point racks and fire at 1500°F for approximately two minutes or until counter enamel is slightly glossy. Remove hot tiles from rack with metal tongs, turn face up on firm, heat-resistant surface, and gently flatten with steel press plate (planche). When tiles are cool, remove excess firescale with toothbrush and straighten out any warpage by gently bending.

Note: Golden area is where liquid flux has been fired.

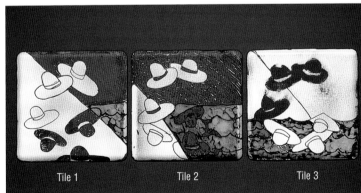

Photo #1

89

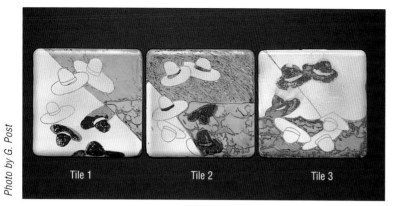

Tile 1   Tile 2   Tile 3

Photo #2

9. Mask off any areas where firescale is to remain. In this piece, mask consists of hastily applied masking tape. Place tiles in pickle solution for approximately 15 minutes to remove unwanted firescale. Rinse in clean water and dry. See Photo #2. Fluxed areas are no longer golden; they have been etched by pickle to a matte surface.

10. Start with medium fusing flux, using 150 mesh sifter, and sift lightly over entire surface. Using 200 mesh sifter, sift transparent beige over top half of image and various transparent blues, turquoises, and purples over bottom half. Sifting should mimic a slight wash of color. Use one enamel at a time over one sheet of catch paper and keep application thin so as not to dislodge particles and contaminate enamels.

11. Place each tile face up on a point rack and fire as in Step 8 on page 89. Fire to orange peel or slightly underfire.

12. When tiles have cooled, file edges to remove oxidation.

13. Place small amounts of 2–3 opaque whites and off-whites, -325 mesh, and a small amount of a darker transparent blue, 80 mesh, in separate compartments of watercolor palette and moisten slightly with water. Use fine synthetic sable watercolor brush to gently push -325 mesh enamels into areas of images that will be lighter in color than background. Push coarse particles of blue transparent into areas needing to be darker (e.g.: the hatbands in Tile 1 above).

Note: *Particles should be wet enough so that paintbrush uses water to carry particles into place.*

14. After enamel has dried, sift thin layer of flux over each tile with 150 mesh sifter. Flux layer serves to even out layer over the newly applied wetwork. It need not be thick because if it were, the layer would take longer to fire and opaques would become more transparent than desired.

15. Repeat Steps 11–12. See Photo #3

16. Repeat Steps 13–15 as many times as necessary to further build contrast and color.

Photo #3

17. Sift a very soft white opaque through 200 mesh sifter over hats in Tile 3 below to further blur images. Brush away excess with soft synthetic sable paintbrush. Fire as in Step 11 on page 90, then repeat Step 12.

Note: There is no need to stencil the area to sift over because sifting is very thin and it can be brushed away without leaving any trace.

18. Sift a soft flux through a 150 mesh sifter over all three tiles. Carefully place red and light turquoise glass balls, with curved tweezers, making certain not to disturb top layer of flux. Fire to gloss. Glass balls should retain their shape.

Finishing:

1. Finish piece by smoothing metal edges and mounting for a wall-hanging triptych. See Photo #4.

Photo #4

Averill B. Shepps

Shepps recognized as a child that she wanted to spend her life in Art or Science. A logical mind and a love of the natural world led her first to the sciences where she distinguished herself by graduating Magna Cum Laude from Smith College with a major in geology and a minor in art. Shortly thereafter she discovered enameling, fell in love with it, and has pursued it ever since. While her science background has enabled her to better understand some of the technical aspects of enameling, the discipline of experimenting and then testing the results to make certain the experiment can be reproduced, which is followed by all scientists, has been even more useful. Interesting effects can be experimented with and reproduced if careful attention is paid.

Shepps has been creating pieces for sale and exhibition for 40 years, having supported herself by selling her work at major arts festivals and craft shows and occasionally teaching classes and workshops. She has developed her own often innovative methods of enameling and works with transparent colors almost exclusively. While she produces plates, bowls, wall pieces, and jewelry, her more important statements are made on larger plates and framed enamels.

Averill

She says of her work, "I want my work to look as free and spontaneous as possible. I hope the viewer will respond to the freedom in the design or the lushness of a brush stroke frozen in enamel as he or she would to a Sumi-e painting." Yet enameling is such an exacting medium that a true sense of freedom in design and execution is almost impossible using traditional techniques. Shepps has developed ways of working that enable her to convey this sense of freedom.

Her project here is one in which she has applied several layers of enamel on the piece before placing the piece in the furnace to fire the enamels. She has found that fewer firings yield an end result that looks fresh and spontaneous, not overworked.

Photo by Anton Shepps

92

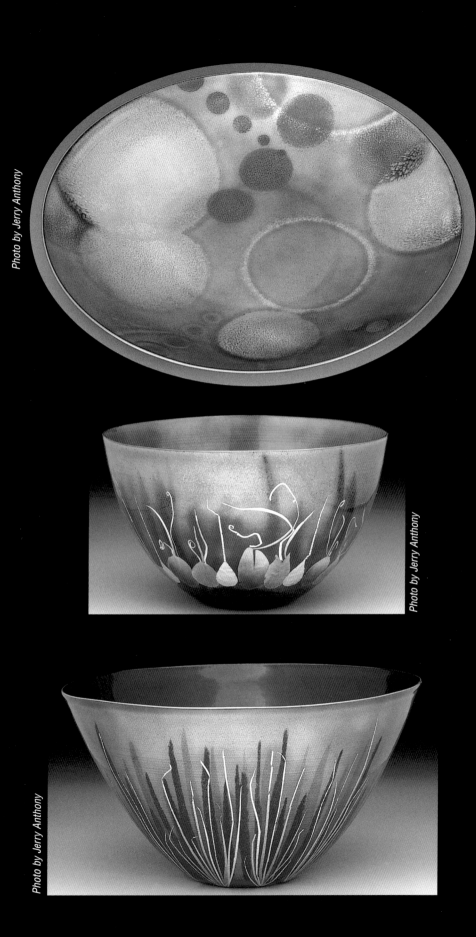

Minimal Firing Enameling

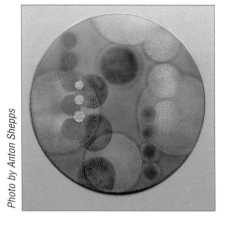

Photo by Anton Shepps

TOOLS:
Circle templates
Firing tools
Paintbrushes
Sifter: 80 mesh
Spray/mist bottle
Trivet that holds copper piece by edges

MATERIALS:
Copper disk: 6"–8"
Enamels: 80 mesh (see chart below)
Klyr-fire®/water solution, 1:4
Manila file folder or other stiff paper
Paper towels

Techniques to Know
- Applying Enamels: Sifting, pg 31
- Preparing Metal for Enameling, pg 27
- Cleaning Between Firings, pg 28
- Working with Foils, pg 36

Enamels	
Name	**Type**
Gray	O
Green	O
Green	T
Orchid	O
Purple	T
Soft flux	T
White	O
White liquid enamel	

This project was designed to make an enameled piece with as few firings as possible. Its success, however, depends on the choice of enamels. Those that result in interesting textures when they are fired over the copper and each other will produce the best results.

Metalwork:

1. Prepare copper disk for enameling. If using an abrasive compound be certain to rub in one direction only.

Enameling:

1. Prepare work area for sifting and fill sifter with white enamel.

2. Using spray bottle filled with 1:4 Klyr-fire/water solution, spray top of piece.

3. Immediately place a circle template over area on piece where circle of white is desired. If needed, protect bare copper by using strips cut from file folder, and sift enamel onto damp copper. When area is covered with enamel so that copper cannot be seen, remove file folder pieces and then template.

4. Spray piece again with Klyr-fire/water solution. Use enough solution to make enamel damp, but not so wet as to cause it to run.

5. Place template in another area of copper and again sift white enamel as in Step 3. Remove, clean, and then dry template. Clean any undesired enamel grains off bare copper with small brush. See Photo #1.

6. Prepare a sifter with medium- to low-firing flux and spray piece with Klyr-fire/water solution until enamel is damp. Place template so that it overlaps some of

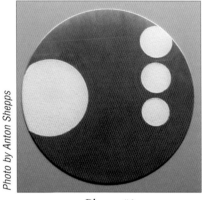

Photo by Anton Shepps

Photo #1

Photo by Anton Shepps

Photo #2

white areas and some of bare copper, then sift enamel as in Step 3 on page 94. If enough Klyr-fire has been used, template can sit on previously sifted enamel without disturbing it. Template may then be placed in other areas of design and enamel sifted onto piece as in Step 5. See Photo #2 on page 94.

7. Clean any undesired enamel grains from bare copper and spray piece with the Klyr-fire/water solution.

8. Prepare a sifter with transparent enamel.

9. Using colored enamels, repeat process of placing the template over areas where color is wanted, then sift and spray as done with white and flux. See Photo #3.

10. When there is enough color on piece, apply cover coat of enamel. To do this, prepare sifter with a pale transparent color that will remain transparent when fired directly over copper.

11. Spray entire piece once again with Klyr-fire/water solution and sift final coat enamel over entire surface. Spray and sift until there is no more copper showing through final coat of enamel. Apply more enamel over bare copper and less over areas that are already covered with enamel. Spray so that enamel is damp throughout. See Photo #4.

12. Place piece upside down on a trivet that holds piece by the edges. To do this, hold piece in one hand, face up, and turn trivet upside down with other hand. Place trivet over piece, then turn both so trivet is right side up.

13. Using liquid enamel, paint underside of piece until it is evenly covered. Be careful not to get liquid enamel on edges or overlapping onto front surface. See Photo #5.

14. Let piece dry, then reverse it on trivet so that front side is up.

15. Fire at 1550°F for two minutes. Firing time may vary depending on colors used. Cool.

16. Clean firescale from exposed copper edges.

17. Repeat process of spraying, placing template to overlap parts of the design, and sifting colors onto fired surface. Clean template between colors. Use opaque colors, choosing those that will give interesting textures as well as color when fired. See Photo #6.

18. Fire at 1500°F–1550°F for two minutes. Exact time and temperature will depend on colors used.

19. If needed, add more colors or accents of gold or silver foil, then fire. See Photo #7.

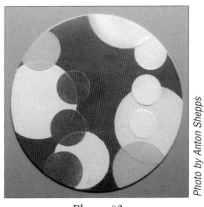

Photo #3

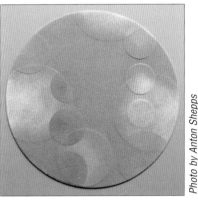

Photo #4

Photo #5

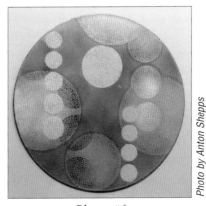

Photo #6

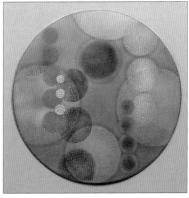

Photo #7

Marian Slepian

"Enameling took me away from other painting mediums, into a realm of glass, metal, and fire—a realm where every step is exciting, every new creation a challenge. The technique of cloisonné particularly entrances me; the play of cool silver lines against the brilliant depths of the glass has a mysteriously sensual and intellectual fascination. The luminous colors are a painter's playground, and shaping the metal is exciting."

For most of Slepian's enameling career, she created large wall hangings and architectural installations. Recently she has found expression in making fine-silver cloisonné objects and vessels for home and liturgical use.

Cloisonné Spice Box
6" x 3"

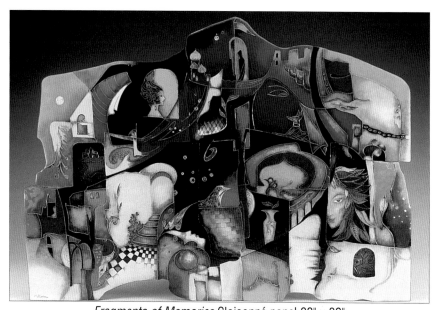

Fragments of Memories Cloisonné panel 22" x 32"

Concerto D'Aranjuez Cloisonné panel 16" x 34"

Mounting Enamels

TOOLS:
Caulking gun
Countersink drill bit, size of stove
  bolt head
Marking pen
Power drill and bits
Power saw
Steel weights (if possible with
  handles)

MATERIALS:
Black matte spray paint
Boards: 2" x 4" x 6" (several)
Bonding agent
Enamel wall piece
Exterior plywood: ⅜" thick
Lag bolts or screws
Metal enamel track 16–18ga steel
Metal wall track 16–18ga steel
Stove bolts and nuts

1. Using power saw, cut exterior plywood to the shape of enamel wall piece, making it slightly smaller all the way around.

2. Mark wall where wall mounting track goes and make a corresponding mark on wood backing. Mounting track should be above center on enamel piece so it will not tilt forward.

3. Using countersink drill bit, drill holes on the front side of plywood backing where the enamel piece will be bonded. This ensures enamel piece will lie flat on plywood backing when stove bolts are installed to attach plywood backing to metal track.

4. Spray-paint the plywood black, on all four edges and on the back where track will be mounted. This seals wood and renders it generally invisible.

5. Place bolts into the drilled holes. The number and diameter of stove bolts depends on the size and weight of the enamel piece. For heavy pieces, oversized washers may be used with stove bolts to give greater strength. See Diagram A.

6. Place plywood backing on boards so plywood backing is amply supported and stove bolts can extend through back of plywood. Enough boards should be used so plywood will not bend or buckle when enamel piece is placed on it.

7. Using caulking gun and bonding agent, bond enamel pieces to plywood backing. Many pieces consist of several enamel panels. Each is bonded by applying bonding agent onto wood in appropriate area. Position panel into place and weight with heavy steel block.

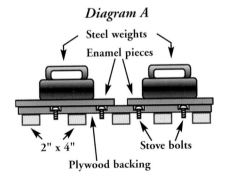

Diagram A
Steel weights
Enamel pieces
2" x 4"
Stove bolts
Plywood backing

Author's Note:
GE Silicon® caulking is the bonding agent I prefer. It is available in all hardware stores. It remains malleable and the enamel can be removed from the plywood backing by sawing with a nylon fishing line. I do not recommend epoxy glue.

8. Do not move for 24 hours. Enamel panels must have complete contact with plywood backing to achieve a good bond. This may be done in stages by gluing a few panels on at a time.

9. Secure mounting track to plywood backing with stove bolt nuts, making certain it is right side up to fit together with wall track. See Diagram B. Metal track is bent per project specifications and made in the shape of a modified "S" from 18ga or 16ga steel. The "S" is modified so it can be flatter than a regular "S" See Diagram C.

10. Mount wall track onto wall with screws or lag bolts as appropriate. Make sure it is in correct direction on plywood backing. See Diagram D.

11. Slide the two tracks together or lift enamel track above wall track and ease tracks together.

The example shown below is a 6'-wide fine-silver cloisonné on copper enamel, mounted in the lobby of Or-Ami Synagogue in Lafayette Hill, PA, for which it was commissioned.

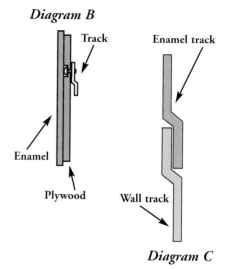

Diagram B

Diagram C

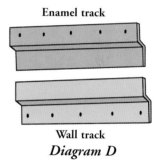

Diagram D

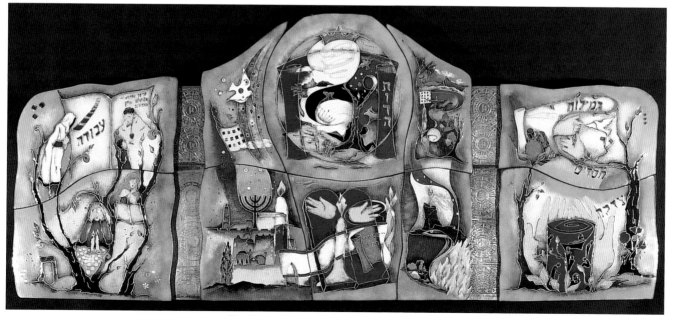

The Three Pillars Cloisonné panel 36" x 72"

Diane Echnoz Almeyda

Diane Almeyda (signature)

Life is never predictable—that's for certain. Almeyda's early life centered around music, and that's where she assumed it would remain. Who would have thought while raising a family, metalsmithing and enameling would step to the forefront?

It all began with a love of jewelry and a habit of scouring pawn shops for something "unusual." This led to a desire for knowledge, not only of stone and metal identification, but also of quality. What better way to judge quality than to learn how something is made. This quest initially led to classes in bead stringing, gem identification, and silversmithing. Things really gained momentum, however, when she was exposed to plique-a-jour enameling at a workshop taught by Valeri Timofeev. It was love at first sight. Challenging, yes; but oh, such beauty in the finished product. Almeyda worked for the next year by herself, learning by trial and error. When she learned that Timofeev's classes in plique-a-jour were being offered at various places, Almeyda made a point of attending them all. Since advanced metalsmithing skills are the basis of plique-a-jour, She continues her studies of new techniques at workshops around the country. Each addition to her repertoire adds an exciting new dimension to her art of choice which she gives back to others by giving classes of her own.

Almeyda's pieces incorporate geometric as well as natural lines and forms, which are complemented by breathtakingly colored enamels. The broad color palette of the enamels plays a large part in the contrast or unity of these elements.

Plique-à-jour Chalice 7⅞" x 3½" x 3½"

Plique-à-jour covered container 2½" x 2¾" x 2¾"

Plique-à-jour bowl 4" x 2¾" x 2"

Plique-à-jour covered box with fine silver and 18k gold

Plique-à-jour Pierced-heart Pendant

Photo by Ralph Gabriner

TOOLS:

Board: 6" sq.
Brass brush or scrubbing pad
Center punch or awl
Clamps
Container: 12-ounce
Dapping block or sandbag
Diamond filing sticks: 200-grit and
    400-grit
Drill bit: #61
Finishing tools
Firing tools
Flexible shaft or hand drill
Glass brush
Jeweler's saw and blades: #2/0
Metal spatula with twisted handle
Needle files: flat, round, etc.
Paintbrush or clay shaper: small
Pickling tools
Plastic containers: small
Sifter: 325 mesh, with top and bottom

MATERIALS:

Dishwashing soap
Enamels: +325 mesh, transparent
    leaded or unleaded avoid very hard
    firing enamels, artist choice
Fine-silver sheet: 1½"sq., 16ga
Klyr-fire®
Rubber cement
Silver chain
Silver split ring
Thumbtacks
Tissue or paper towels

Techniques to Know

• Applying Enamels: Wet packing,
  pg 32
• Basic jewelry skills
• Cleaning Between Firings, pg 28
• Cleaning Enamels, pg 23
• Finishing, pg 37
• Preparing Metal for Enameling,
  pg 27

When doing pierced plique-à-jour, it is helpful to think of the pattern as a stencil. There is a solid framework with cutouts where the enamel will be.

Metalwork:

1. Adhere pattern onto a piece of 16ga fine-silver sheet approximately 1½" square. See Photo #1.

2. Using a center punch or awl, tap a "dimple" in each area to be removed. This will keep drill from slipping when drilling the holes.

3. Drill holes, making certain to keep drill perpendicular to surface. See Photo #2.

4. Using a jeweler's saw, cut out marked areas, making certain to keep saw blade perpendicular to metal and staying just slightly inside lines on pattern. Filing and sanding will take care of any excess metal left inside pattern lines. See Photo #3 on page 89.

Note: Looking at reverse side (the side with no pattern) it will be possible to see if cut-out of design looks right and that opposite sides "mirror" each other.

5. Cut out perimeter heart shape, staying just outside pattern lines.

Photo by Ralph Gabriner

Photo #1

Photo #2

6. Using whatever needle files will fit, file inside edges of holes perfectly perpendicular (at right angles) to top and bottom surfaces of metal. Also file outer edge symmetrical and smooth. Remove pattern.

7. Cut sandpaper into long narrow strips and attach to board with clamp or thumbtacks. Clamp board to bench top so that pressure can be applied to sandpaper strips. (Sandpaper strips can also be held in vise.) Thread sandpaper strips through cut-out holes and sand interior edges smooth and straight. Also sand perimeter symmetrical and smooth.

8. After all holes have been cut, filed, and sanded, sand both top and bottom sides of entire piece so they are perfectly flat and all nicks and dents are removed. Also sand edges smooth. See Photo #4.

9. In a dapping block or sandbag, gently dome the metal. Generally the convex side will be the side people will see when piece is worn.

10. Anneal metal in furnace at 1200°F. Pickle in warm solution. Rinse.

11. Prepare metal for enameling by scrubbing metal with brass brush or plastic scrubbing pad in soapy water, making certain to rinse well. Dry.

Enameling:

1. Clean enamels: grade-sift through a 325 mesh screen. Wash +325 mesh enamels in distilled water until rinse water is clean.

2. Mix 1:5 Klyr-fire/distilled water solution in a container.

3. Pour small amount of solution into plastic cup with wet enamel, swirl, then pour off Klyr-fire/water solution.

4. Again pour small amount of solution into plastic cup with wet enamel. Enamel should be completely covered by a small amount of liquid.

5. Using spatula or paintbrush, scoop a small amount of wet enamel from cup and transfer it to cut-out holes. It is important to have enough water on spatula or brush, as well as enamel powder, as these holes will be filled using the principle of surface tension (like a soap bubble held in a wand). Insert brush or spatula into hole, touch the side, and pull it around so that enamel/liquid mixture spans entire opening. Add more enamel until hole appears to be completely covered. Check back side to be certain solution is not running out, if it is, wick out some liquid and continue. Vibrate piece by rubbing the ribbed side of tool against piece to settle/compact enamel powder. Turn piece over and vibrate again to center enamel in opening. Carefully wick off excess water with a piece of tissue or paper towels.

6. Repeat Step 5 until several holes are filled and powder starts drying. (When previously enameled holes start losing powder when piece is vibrated, it is too dry.) Clean any stray grains of enamel off top of metal with small damp brush or "clay shaper." Transfer piece onto a trivet and place on wire mesh on top of furnace. Let dry completely.

Photo #3

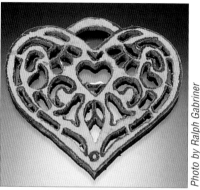
Photo #4

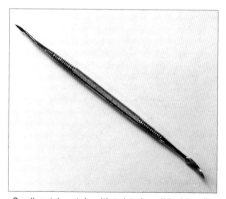
Small metal spatula with twisted or ribbed handle (a modified wax or dental tool works great) or small paintbrush with notches cut along handle (to vibrate enamel mixture)

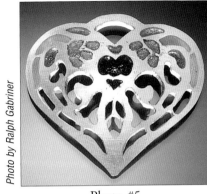

Photo by Ralph Gabriner

Photo #5

7. Fire piece at 1400°F for 30–60 seconds. Exact timing depends on size of kiln, accuracy of pyrometer, how much heat is lost from opening door, etc. Time first-time firing for use with subsequent firings. For first firing "peek" until enamel begins to darken and melt. Do not fire beyond "orange peel"—that is, do not fire to maturity. When enamel is completely melted and the opening is not evenly filled, it will tend to blob up and cling to one side. The enamel must hold together yet still span opening as much as possible. If doing this method properly, any holes should get progressively smaller until enamel covers the entire opening. See Photo #5.

8. Repeat filling and firing sequence, filling both new (unenameled) openings and partially filled cells until all holes are completely spanned by enamel. When there is a complete covering of enamel in each opening, fire piece for 75 seconds to mature and smooth out enamel. Everything should come out filled and smooth. If some of the cells have new "holes" in them, add enamel as before and fire to orange peel until cells appear completely spanned again. Once more, do a "final" firing of approximately 75 seconds. When all holes are filled the piece is ready for finishing.

Note: There is no "magic" number of times for firing. This project had approximately 20 firings. Just do "whatever it takes."

Finishing:

The goal is to remove stray enamel from metal surface, raise a gloss on the enamel, and polish the metal.

1. Using 200-grit diamond filing stick under running water, remove excess enamel from top of piece. Progress to 400-grit.

2. Using finishing papers, remove stray enamel from back side of piece.

3. Use 400-grit finishing paper on a sanding stick to smooth and even out the front. Progress to 600-grit.

4. Wash well under running water. With a glass brush and soap remove metal burs and sanding debris. Rinse well and dry.

5. If any cracks have developed, add a few grains of appropriate colored enamel. Be certain to keep metal surface clean of stray grains of enamel.

6. Place piece on trivet on wire mesh and thoroughly dry on top of furnace as in Enameling: Step 6 on page 103. Fire at 1400°F for 75 seconds. This should give enamel a nice shiny smooth finish.

7. Metal can be highly polished or left with a sanded matte finish.

8. Add a purchased split ring to the upper loop (or make a jump ring) and insert chain. As a bonus, when not wearing the plique-à-jour pendant, hang it in a sunny window and enjoy its beauty.

Photo by Ralph Gabriner

Enameling Photo Gallery

Examples of Fine Enamel Pieces from Professional Enamel Artists

Diane Rooke-Harris

Dream Catcher Enamel on folded copper

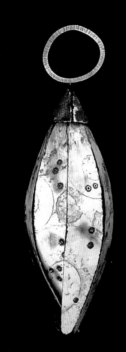

Christina T. Miller

Vitreous Variations

W. Doris Ratz

Marureen Cole

Feather Blues

Beyond the Wall IV

Helen Elliott   *Photo by Kathleen Browne*

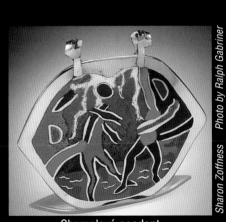

Champlevé pendant

Sharon Zofness   *Photo by Ralph Gabriner*

Jean Tudor

Tudor lives in Edgewood, Washington, having returned to the Northwest after residing in the West, East, the Midwest, and Latin America. She is married to William E. Tudor, a retired Episcopal priest. Her work has been exhibited in England, Germany, Spain, France, Mexico, and the U.S.A. Teaching has taken her to various places in the U.S.A., Canada, England, and Venezuela.

Tudor's formal education was in history, literature, and education. She finally studied enameling during several summer sessions at Penland School of Crafts after great frustration, realizing that there was more to the craft than the shake-and-bake enameling she was doing. Instructors included Mary Ellen McDermott, Fern Cole, Mel Someroski, and Bill Harper. The next major frustration, that of having the enameling ability outstrip what she knew of metals and art in general, was alleviated by getting a BFA in metalsmithing under the tutelage of Phillip Fike and Gene Pijanowski at Wayne State University.

"My involvement with enameling has been stimulated by a number of compelling issues, and I see the path I am taking as a progression through these issues. A combination of the beauty of the materials and the color of the finished product is a basic impetus. Continuing education, for myself and others, about my field is another. Ideas that I want to express is a third. One driving force dominates for awhile, then another takes over."

My Life/Her Life: Ana Cloisonné enamel on copper

Though Tudor continually experiments with materials and processes, she occasionally does some production work. At this point in her work, Tudor is happiest working with ideas and has worked on series involving the psychology of fairy tales, problems of being a clergy wife, and contrasts between her life in Colombia and the life of Colombian women she knew.

"So, these are some compelling forces in my artist life: the beauty of the medium of enamels, the need for knowing as much as I can about my craft, the need to teach and pass to others what I have learned, but above all at this point, the need to express clearly with enamels ideas that surface in my head and life."

Cloisonné over raku background on copper

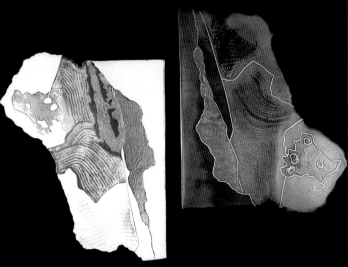

My Life/Her Life: Didsalina Cloisonné enamel on copper

Winter Patterns, Summer Contours Cloisonné enamel on copper

Raku-fired Bowl

Photo by Ralph Gabriner

Techniques to Know
- Applying Enamels: Sifting, pg 31
- Cleaning Between Firing, pg 28
- Counter Enamel, pg 35
- Preparing Metal for Enameling, pg 27

TOOLS:
Finishing Tools
Firing tools
Paintbrushes
Pickling tools or a steel wire wheel
  for removing firescale
Pliers or tongs
Rubber gloves
Sifter: 80 or 100 mesh
Small metal garbage can with tight lid
Spray/mist bottle

MATERIALS:
Copper preformed bowl
Dried leaves, sawdust, pine needles or
  shredded paper
Enamels: Thompson lead-free 80 mesh,
  1995 black (8 oz.), 2430 beryl green,
  (2 oz.)
Klyr-fire®
Newspaper
Scalex
Silver nitrate crystals: pinch

Raku-firing enamel is similar to raku-firing clay. The process involves a reduction firing in which the work is embedded in the combustible material. The fire burns up the oxygen within the enclosure (can). The fire then draws out the oxygen in the metallic oxides which have been used to give color to the enamels. As metal in the metallic oxides is drawn to the surface of the piece, it gives a metallic sheen to the glass. The enamelist must know about the oxides in enamel colorants, and which of those react in reduced oxygen firing.

This project requires excellent ventilation. Do not try this method indoors. Smoke alarms will be activated.

Safety Considerations

• Wear rubber gloves when working with silver nitrate.

Note: Photos do not show this.

• Wear heatproof gloves when working with the fire can.

• Wear a half-mask respirator with a pink/yellow chemical cartridge filter.

Metalwork:

1. Prepare copper bowl for enameling.

Enameling:

1. Paint outside of bowl with Scalex. Dry.

2. Evenly paint or mist inside of bowl with Klyr-fire.

3. Using 80 mesh sifter, sift opaque black enamel onto inside of bowl. The bowl must be well covered so that copper cannot be seen shining through enamel, but do not build up thick coats of enamel. If working with a steep-sided bowl, sift the enamel, then mist again, sift and mist, so that the raw enamel will be well adhered onto copper. Sprayer works best if the mist is applied perpendicular to the surface. If aimed slightly to one side or the other, the force of the spray can blow some of the enamel off, resulting in clumps; however, this will generally smooth out when fired. Dry.

Note: Use an appropriate trivet for large bowl, which is an adjustable wing trivet having wide wings. Nestle bowl upside down within wings, so only outer rim touches. Alternately, rest top lip of bowl on clean wire screen.

4. Place on trivet, fire piece this size at 1550°F–1600°F, then cool.

5. Clean loose Scalex off outside of bowl by brushing it with the hand. Clean edges and outside of bowl.

6. Evenly paint or mist outside of bowl with Klyr-fire, but not so thickly that it runs.

7. Sift opaque black enamel onto the outside surface. Cover so that copper cannot be seen shining through enamel. Dry.

8. Place bowl onto trivet, trying to rest the piece on edges of the bowl. Fire again at 1550°F–1600°F.

Remove from furnace when enamel is shiny. Cool. File edge to remove firescale. If there is a good smooth covering of enamel both inside and out, then continue without adding a second coat.

Note: Do not build up thick layers of enamel, otherwise in the final raku-firing, combustible material will create excessive marks into the surface.

9. Make a newspaper stencil for applying a design. For this project, newspapers were torn with ragged edges and overlaid onto the surface of the bowl. Keep the stencil design simple because an additional texture will be "stamped" into the soft hot enamel by the combustible material during firing.

10. Have both 2430 beryl green and 1995 black enamels ready in sifters so holding agent won't dry before enamel is sifted onto bowl. Paint bowl in areas for stenciling with Klyr-fire. Place stencil onto Klyr-fire. Paint with Klyr-fire again over exposed bowl section to be stenciled. Sift 1995 black onto Klyr-fire. Turn bowl, stencil side down, and pull newspaper stencil down and away from bowl so any loose enamel falls away from the bowl instead of back onto it. See Photo #1.

11. Wearing rubber gloves, mix a "pinch" of silver nitrate crystals with water to make about ½ teaspoon of liquid. This solution deteriorates so don't mix more than will immediately be used. If exposed, this solution stains the skin. Apply solution with brush.

12. Using brush, touch solution into raw enamel stenciled on bowl. See Photo #2. Do not apply so much that it runs down the bowl. At least partially dry bowl to reduce amount of liquid on bowl surface.

13. Lightly spray bowl with Klyr-fire, then lightly brush silver nitrate on shiny enamel. Clean brush in plain water. See Photo #3.

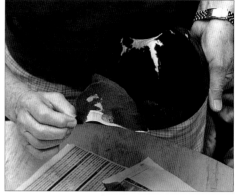
Photo #1

Photo #2

Photo #3

Photo #4

Note: *Silver nitrate applied into raw enamel results in some cratering, sometimes creating a nice rough surface, whereas when applied to fired enamel it makes a smoother surface.*

14. While bowl is still wet, find a few unstenciled areas and sift those areas with a thin coat of 2430 beryl green enamel. Dry thoroughly.

Raku-firing:

1. Have garbage can ready, full of whatever combustible material has been chosen, with a little left over at one side within easy reach.

2. Place bowl upside down on trivet and fire at 1500°F. Sometimes with silver nitrate, the solution will run through cracks which occur when copper expands and the enamel cracks in the furnace. This happens if the temperature is a little low on this firing and/or if the bowl is cool—if not cold—when placed into furnace. A very pleasing hairline pattern results. This effect, though pleasing, is not predictable.

3. With fork, remove bowl from furnace. Try to remove bowl and leave trivet in kiln to avoid extra gouges in the soft enamel. See Photo #4.

Photo #5

4. Flip bowl, pattern side down, into garbage can. Throw a little combustible material on top and wait for material to flame up. See Photo #5. Cover can tightly. If working inside, drag can outside.

5. Leave can covered until smoke slowly seeps out of the can and does not look as if it is being driven by heat. Remove lid and lift out bowl, using heatproof gloves and tongs. See Photo #6. Cool bowl quickly by waving it back and forth in the air. Do not immerse it in water!

Photo #6

110

Finishing:

1. When cool, brush off any sticking burned material. Wash with water and glass brush. Bubbled silver should show where silver nitrate was placed on raw enamel. Smooth silver ranging from silver to silver-blue should show where solution was painted on fired enamel. Iridescent, reddish copper glints will show where the transparent beryl enamel was sifted on. See Photo #7.

Photo by Ralph Gabriner

Photo #7

Author's Note:

Since I use mostly Thompson enamels, the enamels I can recommend for raku-fired projects are from the Thompson lead-free palette.

Transparent colors let go of their oxides best. Opaques do not. Use opaques as "constants," something that will not change drastically and thus be used to control the design. The opaques may pick up some grayish fuming from the smoke, especially the whites and lighter colors.

Copper-oxide enamels work the best; #2435 turquoise, #2430 beryl, #2335 peacock, and #22305 nile are examples of the copper oxide colors. Manganese-oxide enamels are the next best type of color to use. Some examples are #2190 chestnut, #2140 russet, #2715 rose purple, and #2910 oil gray. The manganese colors show as wispy gray or yellowish gray. The color produced is less than the copper oxides, but at times the surface gets a crystallized look.

Some nice combinations to try include #2110 ivory beige with one of the copper oxides or #1020 white with silver nitrate.

Tom Ellis

"I feel very blessed to have enameling on metal in my life. I have always been stimulated by color. Working and manipulating color becomes a means of self-expression. Then there is glass with its power to transmit light—which makes color come alive as light passes through. Add to this precious and semiprecious metals and their ability to reflect light back through glass. This combination of materials creates a visual depth of color and light unlike any other material."

Ellis likes the immediacy of the firing process. The variety of materials, the techniques, and the seemingly infinite number of ways that enamels are used continually amaze him. The ancientness of the medium makes him think about all those who have enameled before. Ellis feels akin to these people as well as those who are enameling today.

"I am attracted and fascinated by this process for many reasons. There are some that I don't yet even know about, for enameling has been and still is a journey of discovery."

Ellis began his enameling career at the John C. Campbell Folk School in Brasstown, North Carolina, in 1975. He was resident enamelist from 1976 to 1981. From 1985 to present, he has been employed with Thompson Enamel in Bellevue, Kentucky, as *Glass on Metal* editor and technical consultant.

Ellis has taught numerous workshops across the United States, Venezuela, Trinidad, Australia, and Canada.

Tom Ellis

Photo by Ralph Gabriner

Separation Enameling

Techniques to Know

- Applying Enamels - Sifting, pg 31
- Cleaning Between Firings, pg 28
- Counter Enamel, pg 35
- Preparing Metal for Enameling, pg 27
- Sgraffito-See project pg 118

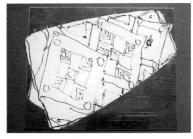

Photo #1

Photo #2

Photo #3

TOOLS:
Finishing tools
Firing tools
Forming pliers
Glass slab
Hammer
Metal shears
Mica sheet
Paintbrush: small
Palette knife
Pencil
Sandpaper 300-grit
Scissors
Sifters: 40 mesh, large and small
Spray/mist bottle
Steel wool: 000, no soap

MATERIALS:
Copper sheet: 18ga, large enough for the design
Copper wire: 2', 12ga or 14ga
Glass chunks: lump form enamel
Klyr-fire®
Paper for design
Rubber cement
Thompson lead-free enamels: 80 mesh, 2110 ivory yellow, 2520 aqua blue, 1030 foundation white, 2530 water blue, SE-2, dry powder, separation enamel

When separation enamel, an accessory product, is fired onto an enamel surface, a depression or indentation develops where the separation enamel was applied. When two or more parallel lines of separation are fired into a transparent enamel, the areas in between the lines become more intense in color and the indented areas less intense in color. When parallel lines are fired into two or three layers of enamel, trace amounts of each color layer become visible. By varying elements to the process, such as color choice, combining opaque and transparent layers, placement of enamels, placement of separation, enamel thickness, and firing time and temperature, many different results are possible. With experimentation, separation enamel becomes very controllable and offers much potential as an enameling technique.

Metalwork:

1. Draw rough sketch of decorative design. Cut out design and glue onto 18ga copper sheet. See Photo #1.

2. Using metal shears, cut out shape and stone edges smooth. See Photo #2.

3. Anneal copper in furnace at 1450°F–1500°F.

4. Hammer copper into shape to form a shallow dish. See Photo #3. Anneal again if metal becomes too hard to form.

5. Pickle piece to remove firescale. Rinse in water and then thoroughly glass-brush the front. Rinse again and dry immediately.

Enameling:

1. Sift transparent 2110 ivory yellow as first coat on front side. Dry and fire at 1500°F for approximately three minutes. Cool.

2. Pickle, glass-brush back side, rinse and dry.

3. Sift 2520 aqua blue on back, dry, and fire at 1500°F for three minutes. After cooling, stone or file edges to remove firescale.

4. Shape copper wire into coils with forming pliers. See Photo #4. These will be fired onto back side of the tray, functioning as both decorative elements and self trivet, eliminating stilt marks from trivet edges.

Photo #4

5. Sift second coat of 2520 aqua blue on back. Carefully place coil into position onto coat of unfired enamel. Dry piece, then fire at 1450°F for three minutes. Cool.

Note: *From this step on, whenever piece is fired and cooled, stone or file edges and then submerse into pickle to remove firescale from coil; rinse and dry. See Photo #5.*

Photo #5

6. Spray Klyr-fire onto front and sift on light coat of 1030 foundation white. Spray again and sift on another light coat. Spray again. Enough Klyr-fire should be used so that enamel is crust-like, not powdery when dry. Dry.

7. Sgraffito basic shapes of decoration into white enamel, exposing yellow underneath. See Photo #6.

Photo #6

8. At this point, piece could be fired, but instead, spray with Klyr-fire and double-sift 1030 foundation white, lightly covering sgraffitoed design. See Photo #7. Fire at 1450°F for 2½ minutes or to gloss. Cool.

9. Spray Klyr-fire and sift medium coat of 2520 aqua blue on entire front. Spray again and using small sifter, sift 2530 water blue on outside diamond shapes. Use paper stencils as needed to block areas. Dry. Fire at 1450°F for 2½ minutes. Cool. See Photo #8.

Photo #7

10. Using palette knife on a flat piece of glass, mix SE-2 separation enamel in a 1:1 Klyr-fire/water solution. Mix to smooth, creamy consistency. Using small brush, apply separation enamel to design. Painted lines should not be much wider than ⅛". When painting lines, the application should be rich enough so that lines are a consistent ochre color when dry. More Klyr-fire/water solution may need to be added during painting process as enamel may dry out and become too thick. Separation is most effective when painted lines of design are at least ¼" from one another, running parallel (or concentric circles, squares, etc.). The idea is that when two parallel lines are fired, enamel flow in between lines is part of the excitement of separation. See Photo #9. Fire at 1450°F–1500°F for 3½–5 minutes, depending on size of piece.

Photo #8

11. Place glass chunks onto mica and fire at 1250°F for 7–10 minutes or until spherical. Cool.

12. Attach spheres to tray's surface with Klyr-fire. Fire at 1450°F for two minutes. Cool.

Finishing:

1. Clean up edges and coils on back. Sand, then buff with 000 steel wool. See Photo #10.

Photo by Ralph Gabriner

Photo #10

Photo #9

115

Sally Wright

Wright began her art education as a child, studying drawing and painting at local museums and art schools. She has always been involved in creating art. She earned her degree in fine arts and a degree in anthropology at the University of Colorado. In addition, Wright studied painting at the Rhode Island School of Design and illustration at the Parsons School of Design. Having been a painter and illustrator by profession, she decided to enhance her artistic base by studying enameling and jewelry making. She has had her own art and design business for several years, and teaches enameling as well.

"I became fascinated with enameling as another medium in which to express myself. The unique qualities inherent in enameling challenge and inspire me as an artist. While enjoying the technical discipline necessary to execute work in traditional methods, I am captivated by the pursuit of expanding and working past those boundaries to achieve a synthesis of my work as a painter and my work as an enamelist. Color, composition, and design are of extreme importance to my work, as are my love of nature and whimsy."

swright

Ocean Eyes Champlevé

Swright

Frog Necklace

Convergence

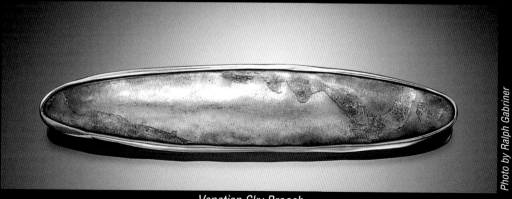

Venetian Sky Brooch

Sgraffito Plate

TOOLS:
Enamel spatula
Enameling tools
File: ½" round
Finishing tools
Firing tools
Scribe
Sifters: 80 and 100 mesh
Paintbrushes: 000, 00, natural bristle
Pickling tools
Plastic scrubbing pad
Spray/mist bottle

MATERIALS:
Copper dish: 4¼" diameter, 18ga
Klyr-fire®
Fine-silver foil
Paint thinner
Paper towels
Powdered kitchen cleanser
Thompson lead-free enamels: 80 mesh
  1540 wedgwood blue, 1422 aqua
  marine green, 2520 aqua blue, 1150
  woodrow brown, flux, 1995 black,
  2020 flux,
Underglaze, black

Techniques to Know
- Applying Enamels: Sifting, pg 31
- Cleaning Between Firing, pg 28
- Counter Enamel, pg 35
- Finishing, pg 37
- Preparing Metal for Enameling, pg 27
- Working with Foils, pg 36

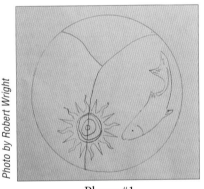

Photo #1

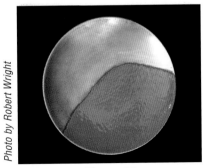

Photo #2

This project was designed to show several ways in which sgraffito might be used—over a base enamel coat, over bare copper, over firescale, over enamel-coated foil, and through black underglaze.

Note: *Eight different plates were created—each one made to show a specific part of the process. Subtle differences in these plates will be noticed.*

Metalwork:

1. Clean a shallow copper plate. Make certain edges are smoothed and filed.

Enameling:

1. Refer to design for plate. See Photo #1.

2. Paint a portion of the plate, corresponding to the right side of the drawing with Klyr-fire. Using 1540 wedgwood blue, sift onto wet Klyr-fire, using medium size 80 mesh sifter. Use a dry, clean brush to clean up and refine edges of this portion of the enamel. Place plate on firing mesh and allow to dry on top of warm furnace.

3. Fire plate at 1500°F until glossy. Underfiring is acceptable, because of subsequent firings. Cool, then stone edges. See Photo #2.

4. Clean off firescale.

5. Paint both "water" portions of the plate corresponding to Photo #1 on page 118 with Klyr-fire and sift with 100 mesh sifter, 1422 aqua marine green. Do not apply too thickly because sgraffito will be done through this layer. See Photo #3.

Photo #3

6. After area is covered with enamel, hold plate at arm's length and using a sprayer with a 1:1 Klyr-fire/water solution, spray just above plate and let mist fall onto enamel. Spray approximately three times. See Photo #4.

7. Place plate on mesh and allow to dry completely on top of furnace.

8. After plate is absolutely dry, gently carry to work station. Using metal or wooden scribe, sgraffito through unfired enamel coat. Sgraffito is a loose technique—exactness is not the goal. A hand-drawn feel is what is sought. Patience and time are important for this technique. Slowly sgraffito through the powdered enamel to reveal surface below. See Photo #5.

Photo #4

9. Gently tilt plate to remove excess enamel material and push it to edges with drawing implement. If too much enamel has been sifted onto surface, the lines may collapse in on themselves—if this is the case, brush off unfired enamel layer and start over. Notice design reveals two surfaces underneath—half the design is over fired blue base coat, and the other half is over unfired copper. Once the sun, fish, and waves are drawn, again hold piece at arm's length and spray it with Klyr-fire/water solution as in Step 6. This spraying helps hold lines intact. Place plate on mesh and allow to dry. When plate is dry, fire and cool. Stone edges to remove firescale.

Experimentation:

1. The piece has been sgraffitoed and fired. To see other sgraffito possibilities, experiment with sgraffito on the layer of firescale in the area of the plate that has not yet been enameled. See Photo #6.

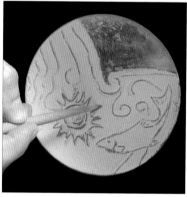

Photo #5

2. Using sharp metal scribe, scratch into firescale surface. Start with circles and lines as they are easiest to control and give an interesting effect. Try different sizes of scribes and shapes. See Photo #7.

3. Once design is scribed, use a cotton swab dipped in pickle solution to lightly swab exposed copper lines in the firescale, then rinse off with water and pat dry with paper towel. Paint entire piece with thin coat of Klyr-fire and sift a layer of unleaded 2020 flux over entire piece. Let dry, fire, cool, then stone edges of the plate.

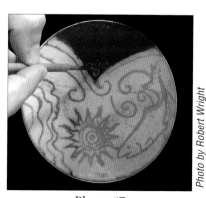

Photo #7

Photo #6

Photo #8

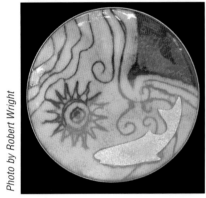

Photo #9

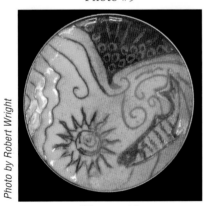

Photo #10

Note: Piece is counter-enameled at the end of the project because it is a domed shape and will not warp as a flat piece would do.

4. The plate could now be considered finished if a coat of counter enamel was placed on the back side. See Photo #8.

5. To experiment further, draw a fish design like the one sgraffitoed on the plate. Use this pattern and cut out a fish from fine-silver foil. To cut, sandwich foil between the drawing and another piece of paper. Paint thin coat of Klyr-fire on sgraffitoed fish, and carefully place foil fish over sgraffitoed one. Use 000 or 00 paintbrush dipped in water or Klyr-fire to pick up foil and transfer it onto plate. Smooth down foil fish and pierce it slightly to avoid air bubbles. Sift a tiny amount of unleaded 2020 flux in a spot adjacent to fish on plate. Fire plate, using flux as indicator for when foil has adhered to enamel surface. Remove plate from furnace, and burnish foil to enamel by lightly smoothing it down with the back of an enamel spatula. Cool and stone edges.

6. Brush thin coat of Klyr-fire onto foil fish and using 80 mesh sifter, sift a small amount of 2520 aqua blue onto the fish. Sift over that with unleaded 2020 flux. Dry, then fire at 1500°F. Cool, then stone edges. See Photo #9.

7. Apply thin coat of Klyr-fire over fish and right side of sun design. Lightly sift 1150 woodrow brown over Klyr-fire. Dry. Sgraffito through brown enamel and leave just enough to accent fish and sun. Fire, cool, then stone edges. See Photo #10.

8. If desired, add some accents to design of fish with a bit of powdered black underglaze mixed with water and a drop of Klyr-fire. Apply with 000 or 00 paintbrush, let dry, then sgraffito excess away. When thoroughly dry, sift unleaded 2020 flux over Black underglaze. Fire plate, cool, then stone edges.

9. Clean back of plate to remove grease from contact with fingers.

Note: Back of plate will be covered in a sort of "gun metal coating" of firescale—not the type that flakes off. This could be removed, but it doesn't matter because it will be covered with an opaque color.

Photo #11

Finishing:

1. Spray or brush Klyr-fire onto plate back. Sift on 80 mesh black opaque enamel. Allow plate to dry thoroughly. Place it on edge, with black back facing up, on mesh and fire at 1500°F. Cool, stone and file edges, then polish with rouge compound. See Photo #11.

Note: Even though piece is made with unleaded enamels, it is not recommended for use with food.

Stenciled Tile

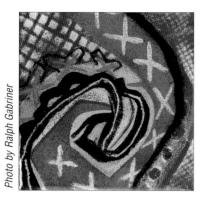

TOOLS:
Enamel spatula
Enameling tools
Finishing tools
File: half round
Firing tools
Paintbrushes: 000, 00, natural bristle
Pickling tools
Plastic scrubbing pad
Scissors: small, sharp
Scribe
Sifter: 80 mesh
Firing mesh: 2 screen sizes

MATERIALS:
Copper: 4" sq., 18ga
Klyr-fire®
Manila file folder paper
Powdered kitchen cleanser
Thompson lead-free enamels:
  1150 woodrow brown, 1315 willow green, 1465 peacock blue, 1319 bitter green, 1124 cork brown, 1422 aqua marine green, 1995 black

This design was made with the intention of creating a simple project using handmade stencils, and simple items found in the studio as stencils. This piece employs only stencils with a bit of sgraffito and inlay work. Obviously, if this technique were used in conjunction with others, or was enhanced with foils, etc., the results would be even more dramatic. See Photo #1 for project drawing.

Techniques to Know
- Applying Enamels: Sifting, pg 31
- Applying Enamels: Wet-packing, pg 32
- Cleaning Between Firing, pg 28
- Counter Enamel, pg 35
- Finishing, page 37
- Preparing Metal for Enameling, pg 27
- Working with Foils, pg 36

Photo #1

Making Stencils:

1. Stencils can be made from many types of paper, plastic, old X rays, metal, etc. For this project, photocopy design onto manila folder paper.

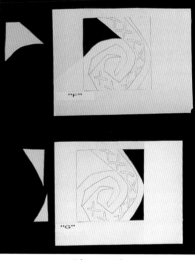

Photo #4

2. Using scissors, cut out stencils. Both the positive and the negative portions of each stencil are kept because both have uses.

Stencils to make for this project: A, B, C, D, E, F, and G. See Photos #2–#4.

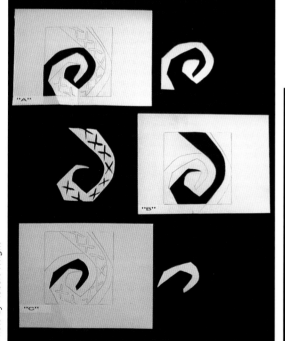

Photo #2

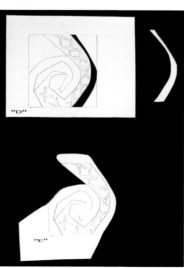

Photo #3

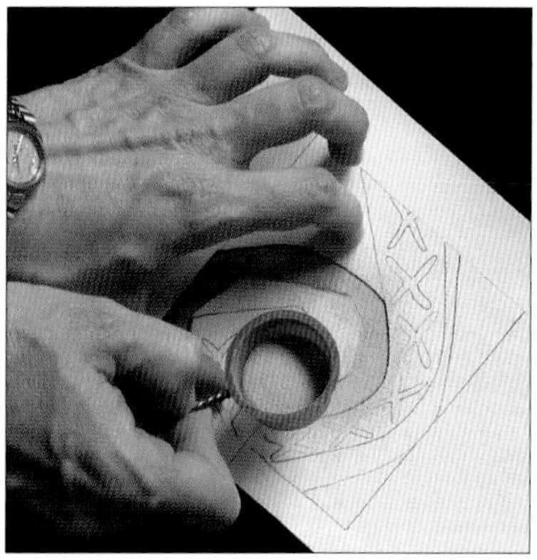
Photo by Robert Wright

Photo #5

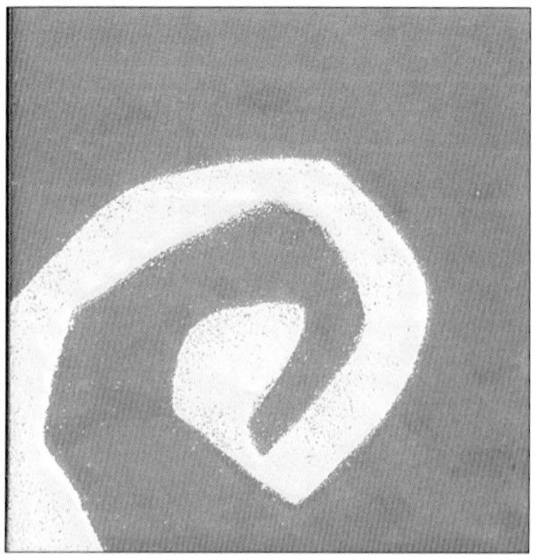
Photo by Robert Wright

Photo #6

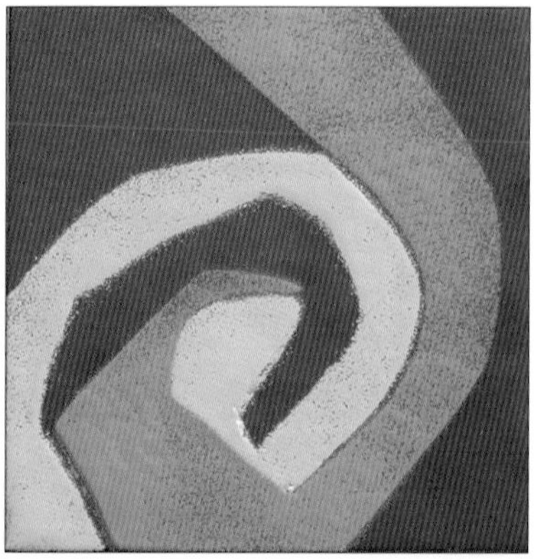
Photo by Robert Wright

Photo #7

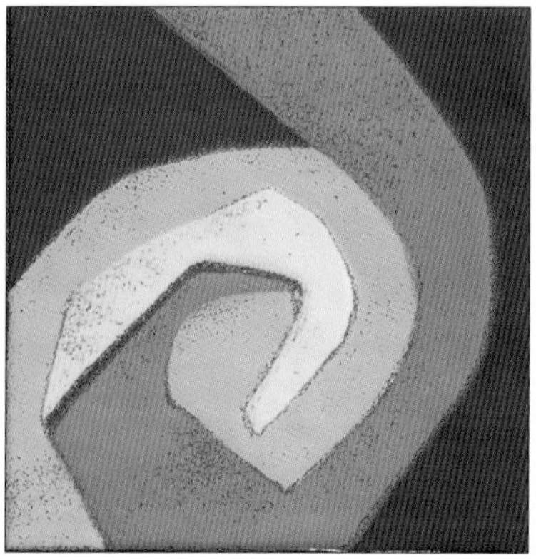
Photo by Robert Wright

Photo #8

Metalwork:

1. File and smooth edges of copper piece. Prepare metal for enameling by including the heating in furnace option.

Enameling:

Note: *Place piece on platform before sifting so piece can be easily lifted and moved without disturbing sifted enamel. Brush klyr-fire on before each stencil is applied.*

1. Use medium size 80 mesh sifter approximately ¾ full of 1150 woodrow brown. Paint an even coat of Klyr-fire on piece. Sift three thin coats instead of one heavy coat. Carefully place piece on stainless steel mesh. Dry thoroughly on top of warm furnace.

2. When piece has dried, fire at 1500°F until surface is glossy. Remove piece from furnace and cool.

3. Clean firescale from piece.

4. Repeat Step 1, sifting counter enamel on back of piece. Place piece on firing trivet and repeat Step 2 to fire and cool. Stone edges, clean, then dry surface.

Note: *Piece should be placed on trivet for each subsequent firing in the project.*

5. Check front of piece and make certain there is an even coat of enamel. If not, repeat Step 1 to apply another coat.

6. Select stencil A and prepare to sift enamel over it by again having sifter ½–¾ full of 1315 willow green. Brush Klyr-fire onto enamel surface. Place stencil on piece and lightly brush Klyr-fire on edges of stencil from which the image is to be transferred. This technique gives crisp edges to image, as does holding stencil flat against metal surface. Holding stencil against piece with one hand, sift enamel with other hand. See Photo #5. Sift a complete coating, but not too thick—a little pull-through of the color can give extra dimension to piece later.

7. Carefully lift stencil up and off piece so enamel is not disturbed. Tap off stencil and remove all enamel residue so stencil may be used again for another project. This is the first clearly defined stenciled image; if it has blurred through lifting stencil away, simply remove excess enamel with small brush. Let piece dry completely and fire at 1500°F until glossy. Underfiring is acceptable because piece will have many subsequent firings. Remove piece from furnace, cool, then stone edges. See Photo #6.

Note: *To make stencils easier to lift, make a "T" shaped handle out of tape and stick it onto the stencil before using.*

8. Apply stencil B as in Step 6. This time, use 1465 peacock blue. Dry, fire, cool, then stone edges. See Photo #7.

9. Apply stencil C as in Step 6. This time, use 1319 bitter green. Dry, fire, cool, then stone edges. See Photo #8 on page 122.

10. Cover area of design already stenciled with stencil E. Using the two wire meshes as stencils, place meshes at different angles on sides of design. See Photo #9. Sift 1124 cork brown over stencils. Lift meshes off piece carefully. Dry, fire, cool, then stone edges. See Photo #10.

11. Place the "X" portion of stencil B2 on piece and sift over area to be covered by stencil B with 1422 aqua marine green. Don't worry about a perfect coating—it should be a bit loose, and color variation is alright. Using right edge of stencil, slightly lift stencil up to give "mistier" quality to piece. Sift over edge, using same color. Lift off stencil and gently sgraffito round shapes through "misty" portion of stenciled image on right. Dry, fire, cool, then stone edges. See Photo #11.

12. Using the "X" stencil again, place in an overlapping position on area covered originally with stencil A. Sift over stencil, using 1465 peacock blue. Dry, fire, cool then stone edges.

13. In the area to be covered by stencil C, sift on coat of 1124 cork brown, then sgraffito through it so 1319 bitter green below it shows through. See Photo #12.

14. Using stencils A and C, sift 1995 black onto piece. Keep coating loose and not too even or heavy. Remove stencils and sgraffito in black areas so that colors below are revealed. Using stencil B2, sift along right side of stencil using 1995 black. Keep it loose and light. Using 000 brush, wet-inlay 1995 black, to which a drop of Klyr-fire has been added, add little accents around "X" stencilwork done in Step 12. Dry, fire, cool, then stone edges. See Photo #13.

Finishing:

1. Piece may be left with a gloss or matte finish. Once piece is finished. It should be mounted or framed.

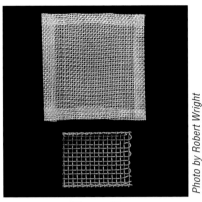

Photo #9

Photo #10

Photo #11

Photo #12

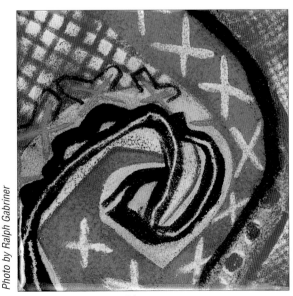

Photo #13

123

Ute Conrad

"The hot kiln has always been my best friend or worst enemy, with a mind of its own," says Conrad, looking back at many years of enameling with a never-quite-predictable firing process. "The temptation to fire just one more time may add the final touch to a piece or ruin it."

Conrad's formal arts training started in the early 1960s at the Academy for Fine Arts in Berlin, Germany. She first enrolled in the Graphics Department. When the holidays approached and money was short, she bought some unglazed china plates and decorated them with porcelain paint to make personalized gifts. Looking for a way to fire them, she went over to the Enamel Department. "This was when my fascination with glass on metal started," recalls Conrad, "so many possibilities, old and new techniques, and the adventure to experiment with color and design."

She graduated with a master's degree and continued as assistant professor for a year. This was an exiting time for the arts in Berlin. She joined a small group of enamelists and craftsmen who experimented with large-scale steel panels for use on buildings, quite a leap for a medium that traditionally had been confined to jewelry and small precious metal objects.

When she came to the U.S. in 1968 to join her husband, who had been assigned to Washington, D.C., she brought her tools and enamel colors along. Conrad established a small workshop in her home in Alexandra, Virginia, and started teaching at the Northern Virginia Art League, the Glen Echo Art Center, and, together with Sylvia Hamers, at the Smithsonian. A few years later, she joined the Enamelists Gallery in Alexandria, one of the few galleries exclusively devoted to the medium, where she still shows and sells her work.

At this time, Conrad's favorite technique is photo silkscreen. "It reminds me of my early graphic work and satisfies my devotion to detail, while color and design remain most important for me." Her inspiration comes from the always present beauty of nature and the fascinating world you discover when you look through a microscope. "This is where the photo process shows its strength," explains Conrad, "I can start my work with an idea and a loan from nature and take it from there." But the photographic image is just the starting point for her work. "Seeing and learning to see has been my goal." She considers her silkscreen work as just one among other techniques. "Enameling offers so much, I will continue to experiment."

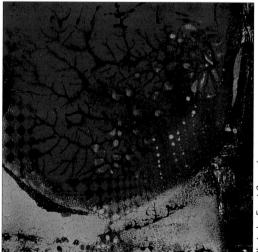

Photo by Ernst Conrad

124

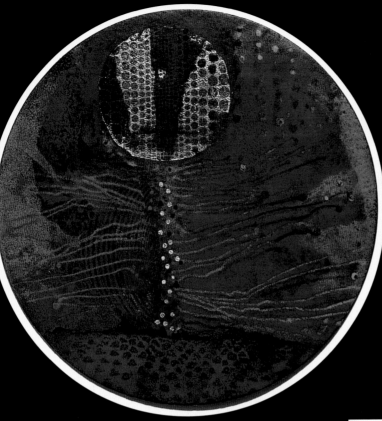

Silver Circle

Silk Screen for Enameling

Photo by Ralph Gabriner

Techniques to Know

- Applying Enamels: Sifting, pg 31
- Cleaning Between Firings, pg 28
- Counter Enamel, pg 35
- Finishing, pg 37
- Preparing Metal For Enameling, pg 27

Photo #1

TOOLS:

Finishing tools
Firing tools
Glass panel, 12" x 16"
Lamp holder with 500-watt photo-flood bulb
Paintbrush: No. 0–8, sable
Pickling tools
Plastic scrubbing pad
Rubber gloves
Sifter: 80 mesh
Spray/mist bottle
Squeegee

MATERIALS:

Black paper or cardboard, 12" x 16" or larger
Cardboard: 3" x 5"
Copper tile: 8" sq., 18ga
Cotton balls
Design: black-and-white photo or drawing, 8" square
Frame with 6xx or 8xx polyester fabric
Photo emulsion kit for silk-screening
Powdered kitchen cleanser
Silver foil
Thompson lead-free enamels: 80 mesh, 1360 jungle green, 1912 nude grey, 1040 quill white, 1915 dove grey, 1319 bitter green, 1840 sunset orange, 1860 flame orange, 1880 chinese red, 2610 sky blue, 2215 soft grey, 2910 elan grey
Transparency film, letter size

The photo silk-screen process allows you to transfer a design onto an enamel tile. Enamel powder is passed through a screen that carries a photographically produced image and works like a stencil. Silk-screening can be used in combination with other creative techniques such as spraying, sifting, and wet-packing.

Note: *The designation 8xx and 6xx refers to weave of multifilament poly fabric. The lower number means fewer threads per inch and larger openings for the enamel to pass through. Enamel in 80 mesh works well with both fabrics. Polyester is recommended because emulsion adheres to it.*

Making the Screen:

1. Create a design on paper and photocopy onto transparency film. A black-and-white design is easiest to work with, but halftones can also be obtained. See Photo #1.

2. Using a wet plastic scrubbing pad and powdered kitchen cleanser, scrub the polyester material in the assembled silk-screen frame. Rinse well and let dry. This will degrease and roughen the screen so the emulsion can better adhere to it.

Note: *Steps 3–5 must be carried out in subdued indirect light.*

3. Evenly coat the screen on both sides with the light-sensitized emulsion, using a squeegee. Let the screen dry for a few hours in darkness. Remember, the emulsion is sensitive to light after mixing with the sensitizer.

> **Author's Note:**
> I use the Diazo Photo Emulsion Kit® by Speed Ball®. It comes with instructions. I use protective gloves when working with chemicals.

4. To copy the design onto the screen, position the transparency over the silk-screen frame, cover with glass panel and use a black background. See Diagram A. To avoid flipping sides of image, place transparency on screen with printed matte side up. Place strong direct light source over transparency. Exposure time for a 500-watt photo flood light, 16" over the screen, is approximately 11–12 minutes.

Note: The emulsion coat must be even and thin. Also the exposure is critical, i.e.: any change in light source, distance, time, or emulsion may cause problems. Underexposed frames will wash out, overexposed emulsion will harden throughout and the image will not wash clear.

5. Under cold running water, use cotton balls to rub screen gently or apply forceful spray of water. Emulsion will wash off, except in areas previously hardened by light exposure. A negative image of your design will appear, i.e.: the black areas in the design will become open on the screen. See Photo #2.

Enameling:

1. To prepare tile for silk-screen printing, clean metal tile, then counter enamel.

2. Clean front side of the tile. Sift 80 mesh opaque or transparent enamel(s) to provide a contrasting background for design. Fire panel. See Photo #3.

Note: Several colors may be used in a pattern that will enhance the design to be printed. Gold or silver foil can also be added at this stage.

3. Place tile under the printing frame. Elevate tile so that it is close to, but does not touch, the screen. Place dry 80 mesh enamel powder on the screen and push through gently by sweeping with cardboard. See Diagram B. No holding agent or water is necessary. Design should now be clearly visible on the tile. See Photo #4.

4. Underfire panel to orange-peel stage so that fine detail is not burned out.

5. Silkscreen process is now complete. Continue working on panel with wet-packing or sifting. Avoiding overfiring.

Finishing:

1. Clean and finish copper edges, and panel is ready for framing. See Photo #5.

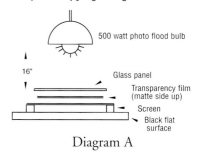
Setup for copying design onto screen

500 watt photo flood bulb

16"

Glass panel
Transparency film (matte side up)
Screen
Black flat surface

Diagram A

Photo #2

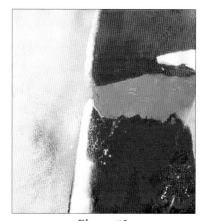
Photo #3

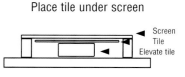
Place tile under screen

Screen
Tile
Elevate tile

Diagram B

Photo #4   Photo #5

Photo by Ralph Gabriner

127

Dee Fontans

The artwork Fontans produces focuses on the female form. She expresses herself through the creation of objects which embellish and celebrate the body. Her work has historically spoken about the human form as subject matter and as a canvas for wearable art. The art is often sculptural, sometimes painterly and is meant to evoke thought and stimulate conversation. The jewelry enriches life through the 'dare to wear' experience. The work on the body pushes and defines the boundaries of space and engages the viewer through aspects of performance. Fontans uses the art as a vehicle to express her social concerns in regard to questions of beauty, ethnicity, and gender.

"I use enamel for paintings of the female nude. The material speaks to me of human history and longevity. Enameling has been a technical vehicle for me to express the beauty of the figure, using the painterly style of grisaille and line drawing. The women in these drawings are round and sensuous, human rather than media constructs. My professional modeling career in life drawing has given me insight in movement through the living form and evoked a passion for the body. These pieces are an investigation into contour line drawing, composition, and material. The enamel brings color into my work. Color is the universal language which expresses mood, life, seasons, and sexuality."

Fontans work speaks to the idea of blurring the lines between art, craft, design, and performance. It deals with conceptual content and technical process.

Dee Fontans

Untitled Nudes Copper, enamel

Photo by Charles Lewton-Brain

Dee Fontans

Kurt Cloisonné pendent

Fran Enamel, fine silver, garnets

Moon Drop Earrings Enamel, fine silver, moonstone

Stone Setting Within an Enamel

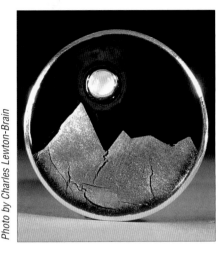

Photo by Charles Lewton-Brain

TOOLS:
Eyedropper
Finishing tools
Firing tools
Flexible shaft with grinding wheel
Glass brush
Jewelry tools
Paintbrush: 00, sable
Pickling tools
Sifter: 150 mesh
Small scissors
Soldering tools
Stir stick
Tweezers

MATERIALS:
Bezel wire: 1.5cm x 4mm fine-silver
Disk: 1¼", 20ga fine-silver
Enamel: black -100/+150 mesh
Epoxy: two-part
Foils: 24k gold and fine-silver
Half jump ring: 18ga
Hard solder
Klyr-fire®
Moonstone: 3mm round cabochon
Pendant finding: ½" x 1", 20ga
   sterling silver

Techniques to Know
- Applying Enamels: Sifting, pg 31
- Applying Enamels: Wet-packing, pg 32
- Basic jewelry skills
- Counter Enamel, pg 35
- Preparing Metal for Enameling, pg 27
- Soldering and Fusing, pg 26

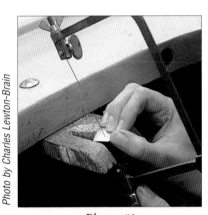

Photo by Charles Lewton-Brain

Photo #1

Photo by Charles Lewton-Brain

Photo #2

To safely work with potentially harmful tools and materials, always use safety glasses, tinted glasses, heavy denim or leather apron, long-sleeved cotton shirt, closed-toe shoes, tied up hair, heat-protecting gloves (leather), dust masks, damp cloths for quick clean up, and when needed rubber gloves.

Metalwork:

1. Fuse ends of bezel wire together.

2. Round bezel. Emery away unwanted marks around outside of bezel.

3. For the metal disk, begin with fine-silver 20ga disk 1⅞" diameter. Cut out if necessary. See Photo #1.

4. File edge of disk to a > shape, a sharp double bevel. See Photo #2. The edge of the > is intended to rest on trivet, making it easier to remove piece from trivet.

5. Clean disk with a glass brush under water.

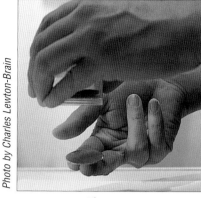

Photo #3

Photo #4

Photo #5

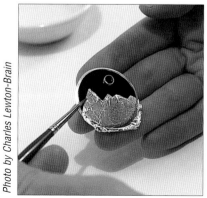

Photo #6

Enameling:

1. Carefully spoon enamel into sifter. Place a coin in sifter to help enamel shake through screen evenly.

2. Brush 1:1 Klyr-fire/water solution onto silver disk. Do not brush solution onto edge trim.

Note: Leave a rim of metal around piece (make a stencil to stop enamel from sifting onto trim). However, if some enamel gets on rim, stone it off after firing.

3. Sift enamel onto silver surface, balancing disk on tips of fingers. Work over a glossy magazine paper. Gently shake sifter, angle metal so that it will catch enamel powder. See Photo #3.

4. Fire at 1500°F. Cool.

5. File edge of disk, maintaining the > shape. File away from enameled area.

6. Disc needs to be sandwiched between two coats of counter enamel and two top coats of enamel. See Photo #4.

7. Place bezel in upper portion of disc.

8. Using 1:1 Klyr-fire/water solution, mix small amount of solution with black enamel and prepare for wet-packing. Enamel need not be "washed" with water first as it will not be seen.

9. Wet-pack enamel half way up inside bezel, taking care not to have enamel climbing up walls of the bezel. See Photo #5.

10. Fire at 1500°F until enamel glosses over. Cool.

11. Check to ensure bezel is secure by gently tugging on it.

12. Keep foil between sheets of protective paper, with design marked on top. Cut paper and gold with small scissors. Place Klyr-fire/water solution on disk where foil for mountains will be laid. Remove top layer of paper and place corner of the foil down over solution. If foil needs adjusting, use wet paintbrush to push foil around enamel disk. Gold foil sheets can overlap. Adding fine-silver foil to enamel inside the bezel will make moonstone appear very white. See Photo #6.

13. Fire and cool.

14. File metal around edge of piece, leaving a fine-silver frame.

Photo #7

Photo #8

Photo #9

Photo #10

Finishing:

1. Bezel should be ⅓ the height of stone. If too high, file down.

2. Set stone, using a bezel pusher, and gently push outside wall of bezel near top edges over the stone, from east and west, north and south, and then all points in between. See Photo #7.

3. Polish setting with tripoli and rouge, using felt buffs on a flex shaft tool.

4. Solder pendant finding to 20ga sterling silver oval (1" x ½"). Pickle and clean with brass brush.

5. Score backs of both oval and pendant loop and enamel disk with a miniature grinding wheel on flex shaft. Work wet to reduce dust. See Photo #8.

6. Scrub finding and enamel disk, using a toothbrush and soapy water. See Photo #9.

7. When preparing two-part epoxy, wear rubber gloves and work in well-ventilated area. Mix even amounts of epoxy hardener and resin until completely blended. Apply glue over scored area of enamel disk. Place finding over top of glue and let it cure for length of time recommended on package. See Photo #10.

Note: *Epoxy bonds best when glued surfaces are pressed tightly together while drying.*

8. Add chain to complete pendant. See Photo #11.

Photo #11

132

Enameling Photo Gallery
Examples of Fine Enamel Pieces from Professional Enamel Artists

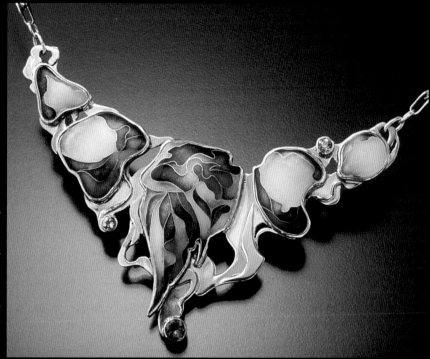

Sheila Beatty   Photo by Allen Bryan

PE8 Necklace

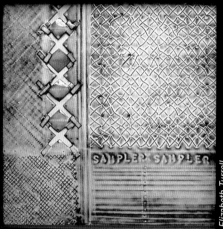

Enamel on stitched copper foil

Elizabeth Turrell

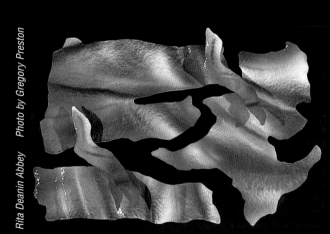

Rita Deanin Abbey   Photo by Gregory Preston

Gondwanaland

Vertebrae Series Collar

Jan Authur Harrell   Photo by Jack Zilker

Roxane Riva

R RIVA

Initially self-taught, Riva sought workshops with a variety of established enamelists. Creative exploration brought her to wall pieces: at first Oriental-influenced landscapes, then, and most notably, to "interior landscapes" and minimalist abstraction.

Riva designed increasingly larger, architecturally inspired pieces in a variety of shapes, firing them at a local enameling plant. Most of her recent work has been these groupings for large spaces, including more than thirty different lobby installations for Sprint and twenty-five reliefs for the Westin Crown Center Hotel in Kansas City.

Her strongest artistic influences are the American Minimalist painters and sculptors for their clean, clear focus, love of color, and contemplative space. Additionally, among enamelist influences are Fred Ball for his creativity, innovation, freedom; and JoAnn Tanzer for her sheer warmth, inspiration, breadth of vision, and technique.

Riva has been active on the governing boards of several Kansas City area arts organizations, has written widely, and was for several years the Kansas City editor for the *New Art Examiner*. She teaches and lectures at all levels, most recently on large scale enameling, enamel collage, and art promotion. Riva's work can be found in glass and enamel museums internationally. She is represented by several galleries and agents.

"For me, enameling is a way to understand my universe. It is questioning, seeking, part of a need to explore. It's a physical love of process as well as an intellectual and aesthetic joy in discovery and application."

The techniques demonstrated in the project are part of a series exploring organic design, shape, texture, and assemblage.

Photo by Gary Rohman

The force that through the green fuse drives the flower

134

Microcosm Series #1

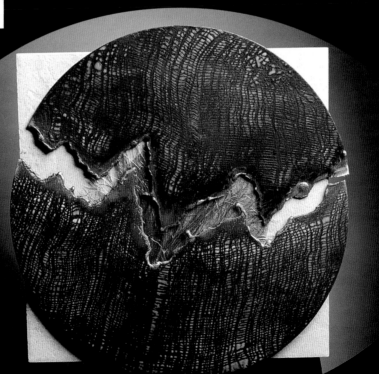

Global Warming: Collateral Damage

Torch-altered Metal with Cheesecloth Stencil

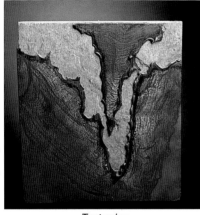

Tectonics

TOOLS:

Cutting area with heat-resistant floor
Firing tools
Heat lamp for drying (optional)
Leather shoes
Natural fabrics (i.e.: cotton, wool, not synthetic)
Oxyacetylene torch, with #5 tip
10# steel press plate (or 2 bricks) to flatten piece and hold during torching
Paintbrushes: assorted
Plastic scrubbing pad
Power drill with wire brush for cleaning, finishing
Sifters: assorted (at least a 10" or 12" sifter for larger pieces)
Spray/mist bottle
Welding helmet with clear (wear over infrared/ultraviolet protective goggles) or green (UV/IR heat and spark protective) shield

MATERIALS:

Cheesecloth: 36" sq. (cotton, not polyester)
Copper sheet: 15" sq., 18ga (or 1" smaller than kiln)
Enamels: 80 mesh, medium- to low-firing, artist's choice
Holding agent
Soft chalk
Thick paper or cardboard for stencil
Vinegar/salt pickling solution

Techniques to Know

- Applying Enamels: Sifting, pg 31
- Cleaning Between Firings, pg 28
- Preparing Metal for Enameling, pg 27

Oxyacetylene torches are available in most sculpture studios and classrooms. Consult person in charge for (A) safety precautions, (B) operating instructions. Be certain to ask for supervision while learning. Plasma cutters are faster and easier, but create a cleaner, "hard edge" cut rather than the fluid, organic line sought in this project.

The oxyacetylene torch alters edges (and interiors) by melting, pooling, and dissolving. Experiment with variables: templates, color, torch tip size, firing time, and temperature. While torching, finishing, or aligning collage elements, be alert to what these elements do naturally, and use these opportunities. Echo exposed copper with more copper: wires, cords, tubing, pipe, perhaps partially enameled. Opaques are generally more successful, providing more definition, but try transparents, too. Think ahead, but always stay open to surprises, accidents, and chance occurrences that may totally change the direction.

Metalwork:

1. Clean copper piece by immersing in 1:20 vinegar/water solution and scrub lightly with salt, brush, or plastic scrubbing pad.

2. Sketch design for shape with soft chalk to mark cutting line. Shallower cuts (closer to the edge) on a single edge will be easier; deeper cuts require more control (both aesthetic and physical).

3. Position copper piece so that area to be cut hangs fully over cutting table edge. This permits melted copper droplets to fall to heat-resistant floor below. These copper scraps can be saved for later collage or ornamentation. Counterbalance copper on supported area to keep piece steady.

4. Cut along chalk marks with torch tuned to a cutting flame (most pointed, blue flame) held at a 90° angle to piece. Hold torch on an edge until metal starts bubbling, and droplets begin to fall to floor below. Slowly follow melted metal with cutting flame through marked pattern. See Photo #1.

Note: If difficult to follow chalk marks through dark glasses or with the flame:

  A. Look ahead of the flame.

  B. Stop as necessary, and lift goggles (or shield) to see the pattern.

  C. Free up! Cut any similar pattern. Cultivate serendipity, the happy chance.

Enameling:

Note: If a larger size piece is a new experience, practice kiln maneuvers with piece on firing support, firing tool, and dark glasses, until your movements are sure and balance is easy.

1. Clean metal as in Metalwork: Step 1 on page 136. See Photo #2.

2. Place metal face down on firing rack or trivet. Brush liquid enamel thickly onto back. Dry. Reverse.

3. From any thick paper or cardboard, cut a freeform stencil for edges (these stencils can be designed for more control). See Photo #3. With sifter, apply base coat of darker, medium-firing 80 mesh enamel. Brush unwanted enamel from masked areas with small acrylic paintbrush.

Note: Avoid high-firing enamels; higher temperatures complicate subsequent cleaning.

4. Place on firing support. Fire at 1475°F for 4½ minutes, or until glass surface is glossy.

Note: Firing times will vary with kiln, size of piece, and thickness. Check from three minutes onward, remembering that each door opening will cool furnace and increase firing time. Avoid overfiring.

5. Remove from furnace. Cool slowly. To ensure flat surface, weight with press plate after seven seconds of cooling. (Be open, however, to the interest of curves caused by warping on a wall piece.)

6. When completely cool to touch (20 minutes or more), reclean piece, using vinegar/salt solution and wire brush or plastic scrubbing pad to loosen firescale from unenameled torched edges. Do not worry if counter enamel is splotchy or uneven. It will serve its purpose unseen.

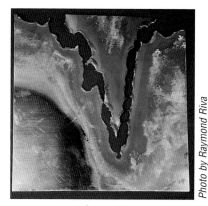

Photo #1

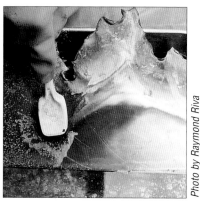

Photo #2

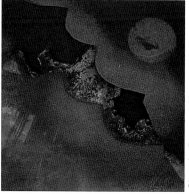

Photo #3

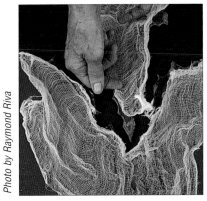

Photo #4

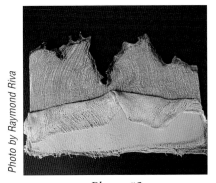

Photo #5

Note: The second coat can alternatively be executed in any other low-or medium-firing enamel technique that takes few firings, e.g.: screens, stencils, painting, wet inlay.

7. Immerse cheesecloth in water; squeeze to remove excess. Brush holding agent onto surface to be enameled. Arrange cheesecloth on enameled piece, pulling and bunching to enhance torch cut lines or thematic intent. Try several patterns. See Photo #4.

Note: If using earth patterns, this technique can look like geologic strata.

8. Remask unenameled areas with stencil. Using large sifter, apply light or medium colored 80 mesh enamels over cheesecloth. Do not overload enamel; medium-light application will enhance delicacy of line and ease cloth removal. Lightly spray cheesecloth with water from atomizer held upright at 18"–24" above piece (this helps adherence of enamel to cloth during removal).

9. Remove stencil from unenameled edges. Gently lift top edge of enamel-loaded cheesecloth, rolling carefully down from top to bottom. See Photo #5. Don't worry if small amounts of enamel fall; blurry areas can be effective. Dry and brush enamel off cheesecloth and stencil for reuse.

10. Let piece dry completely. Fire at 1475°F for approximately four minutes. Watch for first uniform shine of pattern. Remove, weight for flatness if desired. Cool.

Option: For dramatic effect, omit base coat application and firing. See Photo #6. Begin directly at second firing, above, applying cloth and enamel to bare metal. Proceed as directed, but using more enamel, and firing longer or higher (about seven minutes at 1475°F) to fix firescale which forms (matte, black) base. Limit any cleaning to bare edges. Clean as before, this time using wire rotary brush attachment to electric drill. See Photo #7. Finish with 100–300-grit emery cloth. Protect bare copper with lacquer or allow to oxidize.

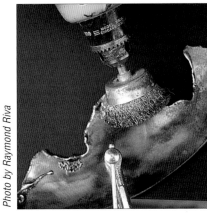

Photo #7

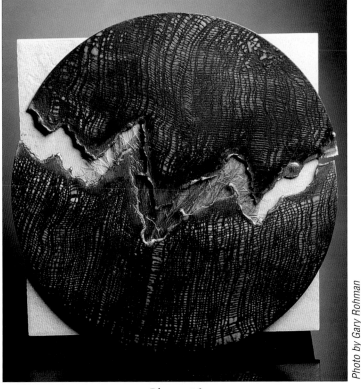

Photo #6

Option: Add collage accents or design elements. Choose among assorted copper pieces, such as copper droplets, "cuts" from torching, or copper wire. See Photo #8.

Option: Curved cuts from and onto 18ga copper bowl forms can be very graceful and three dimensional. Heat color or enamel these for added effects. Beware, however, of loading piece with too many design elements.

11. Attach with any metal-to-ceramic or metal-to-glass adhesive. Be certain to weight wire as it dries so it will not move.

12. Mount piece as you choose. Two simple, yet dramatic mountings:

A. On reverse, glue (flat black painted) wood fitted with hangers at approximately ¾ of piece's height. Add a small block of wood below (at ¼ of piece's height) to assure uniform distance from the wall. Experiment with different wood thicknesses.

B. Attach wood and findings to the back of slate or ceramic tile, and attach enamel to the front. See Photo #10.

Photo #8

Photo #10

Aileen Geddes

AG

Jewelry making started for Geddes over 20 years ago. When she found a piece of wave-tumbled jade at Jade Beach, California, she decided that she would enroll in an adult class given at Carmel. The class was wonderful, Geddes learned the basics of lapidary, free-form stone lapidary, and metalwork, i.e.: lost-wax casting, sheet construction, soldering with propane, and working with an acetylene torch. Within her first year in this class she entered a lost-wax ring that she designed in class and won best of show. Her competition was well-seasoned professionals in Carmel. Needless to say, Geddes was hooked to the jewelry. Since then, she has had many classes and workshops, mostly at Monterey Peninsula College, where she learned hydraulic presswork, basic metal techniques, forging, chain making, lost-wax casting, weaving, new clays, enamels, and cloisonné. With this, she decided it would be great to do beads in enamels.

Geddes went her own way and started experimenting with copper tubes and enamels, until she got a system going with 80 mesh lumps and threads. She had seen a woman using a scrolling wire and actually "painting" designs into her beads. Geddes went home and tried it right away. It worked, and she has been using this method to make beads for the last six years.

Geddes belonged to a gem club in Santa Cruz, at that time. They invited her to demo the bead-painting technique at the gem show. She then participated in the Santa Clara show, which was a federated mineralogical show for the state of California. Geddes next demonstrated for the Northern California Enamel Guild at the California State Fair. She also has demonstrated for the Monterey Gem Club many times as well as giving many private lessons in torch-firing. Geddes loves torch-firing, as no two pieces are the same. It is spontaneous and without rules.

Geddes was an art instructor at Fort Ord, California, for five and a half years. She taught watercolors, etching, pastels, photo silk-screening, advanced mat design, and handmade paper making. She won three major awards for handmade paper scrapbooks at a worldwide competition. The pieces were on exhibit at the Smithsonian in Washington, D.C., and have been shown in many museums throughout the United States.

Geddes feels art is a way for her to share her love for nature and the world around us, whether it be on paper or metal. To her, rocks are the ultimate of natures mysteries and beauty.

Presently Geddes is retired, but she works as a computer artist/illustrator for DLI Presidio in Monterey three days a week. The rest of the time she works on her artwork and goes to shows. Geddes declares she wasn't designed to sit at home and vegetate.

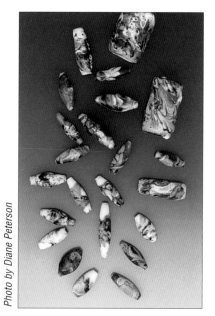

Torch-fired Beads

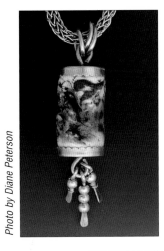

Photo by Diane Peterson

Techniques to Know

- Finishing, pg 37
- Prepare Metal for Enameling, pg 27

TOOLS:

Copper foil squares: 4", 28ga
  to make palette for each color
  of enamel
Finishing tool: 600-grit
Fireproof surface
Glass canning jar
Heatproof casserole lid
Jeweler's saw or pipe cutter
Mask (fume type)
Pickling tools
Propane brazing torch with 4' hose
Safety glasses
Stainless steel rod: ⅛" diameter, 1"
  tapered end to fine point (use as a
  mandrel)
Stainless rod: 26ga, ground to a
  fine point (use as a scroll wire)
Tools for torched-fired nonbead
  objects
Tripod and firing screen
Vise to hold torch while firing

MATERIALS:

Baking soda
Copper or fine-silver tubing (3⁄16"– ⅝"
  diameter, depending on size of
  desired bead)
Disks of same metal as the tube: 22ga,
  size of outer diameter of tube.
Enamel lumps and threads
Enamels: 80 mesh, artist's choice
Hard silver solder (necessary for big
  beads only)
Kiln wash
Klyr-fire®
Propane gas

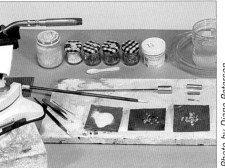

Bead-making tools

Photo by Diane Peterson

Author's Note:
I use a BernzOmatic® torch model #JTH7T with a 4' hose because it is one of the few inexpensive propane torches that gets hot enough to heat the copper and fuse the enamels.

Metal Bead Basics

A wide variety of bead shapes can be made using a torch-firing technique. The instructions here provide steps for a barrel-shaped bead, followed by a short discussion of other shapes.

Use copper or silver tubing. The copper tubing is 3⁄16" and found in hobby stores selling model materials. Begin with the smaller tube until the process is well understood. Purchase larger tubing at hardware stores. It comes in various sizes from ⅛" to ¾". The ¾" size can be used for pendant beads. Install end caps on the larger beads to make them look more finished.

Use propane fuel in the torch because it is a cleaner-burning fuel than oxygen/gas. This brazing torch is used at only half heat, except when firing beads ⅜" and larger. Other types of torches do not put out enough heat to enamel beads.

See Photo #1. The firing mandrel is a ⅛" x 8" long stainless steel rod that is purchased at a welding supply house. One end is ground to a point allowing this one rod to be used for all sizes of beads. See Photo #2.

Leaded enamels work best, especially the lump vitreous enamels as their colors are so vivid. However, nonleaded enamels can be used when the leaded ones are unavailable.

Note: *The larger the bead, the harder it is to fire because the temperature must be maintained between 1450°F and 1550°F.*

Metalwork:

1. Cut tube to desired lengths for beads, use a wooden jig and cut with either a jeweler's saw or a pipe cutter.

2. If tube is larger than ⅛" diameter, cut two disks the size of outer diameter of tube. Drill ⅛"-diameter hole in middle of each disk and solder disk to each end of tube, using hard silver solder. This allows a bead hole small enough to be tightly held on the mandrel. See Photo #2.

3. Deburr tubing with jeweler's file.

4. Clean tubing of any oxides and grease. If using pickle, be certain to neutralize bead in baking soda and water.

Enameling:

1. Cover mandrel tip with kiln wash that has been mixed with water to a creamy consistency. See Photo #1. Place a dab of kiln wash in the hole of the bead. This stops enamel from adhering to mandrel. See Photo #2.

2. Place enamels on copper foil palettes (do not use aluminum foil). Use one 4" copper square for each color/type of enamel. Keep palettes in proper order so it is easy to remember where each color is. See Photo #3.

3. Heat bead in bottom of flame; if bead gets too close to upper flame, it will oxidize and become black. Generally, 3"– 4" from head of torch is appropriate. See Photo #4. Fire in a low-lit room so that color of bead as it heats to proper firing color can be easily seen. When temperature of bead gets about 1500°F degrees it will glow a red to orange color. See Photo #5.

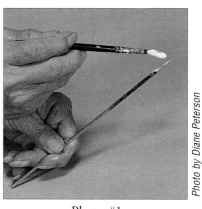

Photo #1

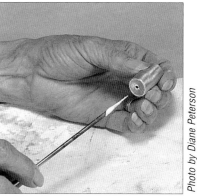

Photo #2

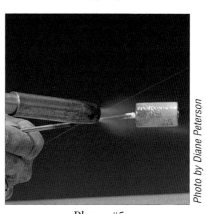

Photo #3

Photo #5

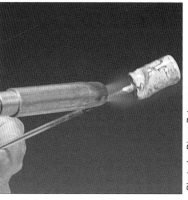

Photo #4

Photo #6

Photo #7

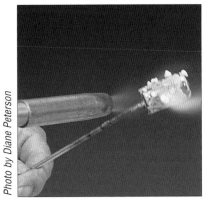

Photo #8

4. When properly heated, roll bead into base-coat enamel on copper foil palette. See Photo #6. If the bead cools down, enamel will not stick. Reheat to orange color again to adhere more enamel; reheat as often as needed, being sure to shape bead as process proceeds.

5. Roll bead at least four times in grain enamel to get a good base coat. Reheat as needed. After 4–5 base coats, roll bead in lump enamels. See Photo #7.

6. After each rolling, heat to maintain temperature, and mold shape as needed.

7. Roll bead in threads.

8. When all design elements are on bead, heat to temperature and use scroll wire as a "brush" to twist and turn the melting glass into a pleasing design. It helps to remember where each enamel color is placed on bead. See Photo #8.

Finishing:

1. Use 600-grit fine wheel, or other fine-grit finishing tool, to clean ends and excess lumps of enamel.

Note: When a set of matching beads is needed, fire them all at one time. This acts like a dye-lot; if they are fired at separate times, it will be difficult to make them match.

Note: Beads can be shaped with the torch by applying heat at ends and tilting mandrel so that enamel flows to center of the bead. See Photo #9. By torch-firing beads, there are many shapes possible, experiment with this technique.

2. When bead is finished, let it cool on heatproof casserole lid.

3. After cooling bead, submerge it and mandrel, into a glass jar filled with water. The kiln wash dissolves and bead drops free. See Photo #10.

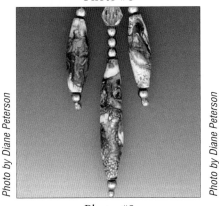

Photo #9

Photo #10

Final bead complete with silver chain and dangles

Enameling Photo Gallery

Examples of Fine Enamel Pieces from Professional Enamel Artists

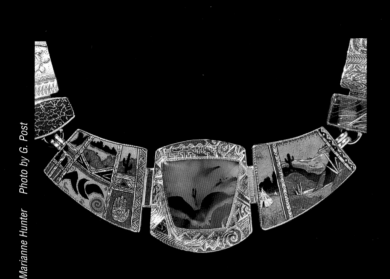

Marianne Hunter   Photo by G. Post

The Desert

Debbie Wetmore

Cloisonné Pins

Karen L. Cohen   Photo by Ralph Gabriner

Cloisonné Purse Mirror

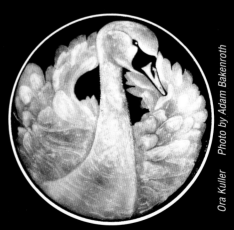

Ora Kuller   Photo by Adam Bakenroth

Grisaille Swan

Sarah Perkins

The makers of hollowware use the properties of the metal—permanence, dimensionality, and plasticity—and the techniques and concern for detail of the jeweler, when creating pieces for people to use together, in public or private ceremonies. Perkins' work has been primarily cup sets and ceremonial vessels because she is interested in the social implications and uses of these forms. The interactions of people with each other and with the objects she makes when they are being used is a conceptual focus of the work. The materials and techniques she uses are chosen because they are appropriate and because they are the ones she is drawn to and enjoys working with.

In her work, the metal forms and the enamel imagery work together to make a whole, with the two materials complementing each other rather than one being visually more important than the other. Technically as well as visually, the pieces are a unit, which is a manifestation of the ideas of integration and relationship that the work addresses. The forms, colors, and imagery in Perkins' work derive from two sources: metal technology and natural forms.

"I feel a strong affinity for metal and I am fascinated by its permanence, malleability, strength, and surface qualities. I use directly or copy in enamel some of the colors and surfaces which naturally occur through the various working processes. It is also important to me that some metal show on the surface, rather than being simply the support mechanism for the enamel, because I want to show that the finished piece is a relationship between and a collaboration of two media.

Classic Folded Vessel Silver, Enamel

"My work reflects my emotional response to my environment, referring to landscape, body part, or natural object. In the jewelry pieces and recent vessels I am concerned with organic imagery. The forms and color imagery are derived from leaves, fish, and other natural objects because I find these modulations of color, texture, and surface both appealing and very intimate."

Relationships between people are also an issue in Perkins' work. The social and ceremonial aspects of drinking, eating, and serving food and drink are referred to with these pieces, and it is therefore important that they be completely functional, whether they are ever actually used or not. The scale is consistent with human use, as are the warmth and softness of the forms and finishes which invite the viewer to investigate the work with both eyes and hands.

Blue Intersections Copper, Enamel, Silver

Jay's Copper, Enamel, Silver

Shares Silver, Enamel, Ebony

Vessel Forms

Photo #1

Photo #2

Photo #3

TOOLS:
Finishing tools
Firing screen and 3-point trivet
  large enough to hold piece.
Firing tools
Metal file
Metal planishing hammer
Metal T-stake
Paintbrushes: 3/0 to 10/0
Pickling tools
Plastic cross-peen hammer
Sifter: 60 mesh
Soldering tools
Sprayer: aerosol or airbrush
Wire wet-packing tool

MATERIALS:
Baking soda
Copper disk: 8" x 20ga; or a prespun
  vessel shape
Enamels: 80 mesh, some opaques and
  some transparents, artist choice
Gold tubing
Gold wire: 18k and 14k round and
  rectangular
IT® solder
Klyr-fire®
Lotus root: ground
Soap

Techniques to Know

• Applying Enamels: Sifting, pg 31
• Applying Enamels: Wet-packing, pg 3
• Cleaning Enamels, pg 23
• Finishing, pg 37
• Preparing Metal for Enameling, pg 27
• Grade-sifting and Particle Sizes, pg 20
• Wirework, pg 29

Metalwork:

1. Begin with a copper disk and raise the desired form on a metal T-stake, using a plastic cross-peen hammer. See Photo #1. Or purchase a spun-copper form available in several shapes. Skip Step 2 if using a spun form.

2. Planish completed form to remove large hammer marks and smooth surface. See Photo #2.

3. Form three cones from gold tubing to make vessel feet.

4. Form square wire to fit top edge of vessel.

5. Solder all metal parts. See Photo #3. Photo shows rim overlapping edge of copper form on both inside and outside. Excess will be removed after enameling. File off excess solder before beginning enameling.

6. Clean away soldering flux with warm soapy water or a warm pickle solution. If pickle solution is used, dip piece in neutralizing bath of baking soda and water, followed by clean water rinse.

Enameling:

1. Mix enamels with 1:3 Klyr-fire/water solution.

2. Wet-pack first coat on inside of form with 80 mesh enamel. Pack as thinly and smoothly as possible without leaving areas where metal shows through enamel. Be certain to pack enamel up under rim. Leave a narrow line of bare metal approximately ⅔ the way up the piece. If the piece is tall, more than one line will need to

be left. These lines of bare metal form a "fire break" to keep enamel at the bottom from pulling areas of enamel off the sides during firing.

3. Fire piece at 1450°F until glossy. See Photo #4. Cool.

Photo #4

4. Wipe off loose firescale, then wet-pack a second layer of enamel on inside. This time cover first "fire break" line(s) while leaving a line of uncovered enamel someplace else. Inside is enameled first because each time piece is fired, the inside enamel will smooth out more and eliminate need to stone inside of piece. Pack area ¼" from rim slightly thinner than on rest of piece because most firings will be done with piece upside down and enamel will naturally slump towards rim, making it very thick.

Note: Two layers of enamel will be enough for inside unless the outside is to be treated with cloisonné wires or some other technique which calls for a thick coat of enamel. Enamel on either side of piece should be about the same thickness.

Photo #5

5. If using transparent enamels, decide if firescale is to be used on outside of piece. If not, clean off firescale before applying enamel. If using opaques, simply wipe off loose firescale before proceeding See Photo #5.

6. Wet-pack enamel thinly using general color families that are related to color selected for finished piece. See Photo #6.

7. Bend gold wire into design for sides of vessel.

Note: Handmade cloisonné wires hammered from square or rectangular wire results in a more expressive line when placed on piece.

Photo #6

8. Adhere wires to curved vessel with strong adhesive such as ground lotus root mixed with water to viscous consistency. Wires can be adhered and fired onto piece approximately ⅓ of surface at a time as adhesive will not hold wires onto the bottom of a curved surface.

9. Fire lightly so wires just stick into enamel surface. See Photo #7. After cooling, pack small amount of enamel around tacked-on wires to ensure they remain in place during subsequent firings when they are rotated to an "underside" position. Scars left by trivet, screen, or other furniture will be mended as more enamel is applied to piece. In this manner, fire all wires into place.

Photo #7

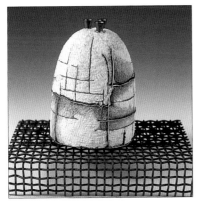

Photo #8

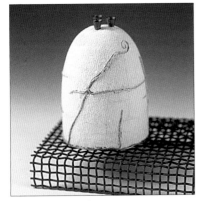

Photo #9

10. Wet-pack surface as with any other cloisonné piece, but use more holding agent than on a flat piece. When firing the bottom half of the vessel, fire it upside down and resting upon the top edge. See Photo #8. Unfired enamel will not stay on any surface tilted at 90° or more. Piece will need to be flipped around on the trivet to maintain a proper angle when enameling the rest of the piece

11. Spray surface with 1:1 Klyr-fire/water solution and sift 80 mesh transparent enamel through a 60 mesh sifter. Brush unfired enamel off cloisonné wires and other metal parts before firing. See Photo #9.

Finishing:

1. When enameling is completed, file and/or sand excess metal off rim. Gently file or grind off excess cloisonné wires as well. Place final finish on inside of rim and fire for a last time to heal sanding and grinding scars.

2. Using 150-grit alundum stone followed by 220-grit stone, smooth surface carefully. Using wet/dry sandpaper up to 600-grit, smooth surface to a satin finish. See Photo #10.

Photo by Ralph Gabriner

Photo #10

150

Enameling Photo Gallery

Examples of Fine Enamel Pieces from Professional Enamel Artists

Audrey B. Komrad

Floral Fantasy

Irene McGuckin & Susan Elizabeth Wood Photo by Hap Sakwa

The Roads Not Taken

Mary S. Reynolds   Owned by Ramona Abernathy-Paine   Photo by Richard Brunck

Palmetto Fan Pendant and earrings

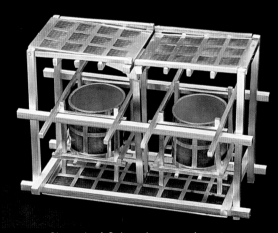

Linda Darty

Champlevé Salt and pepper decanter

Enameling Troubleshooting

There are always problems while enameling. Many mistakes are fixable, but it is also important to learn what not to do in the first place. Throughout the Studio Basics, Tips and Tricks chapter on page 18, notes and tips have been given to try to steer the enamelist in the right direction. Below is a list of problems, possible causes, and possible treatments that can help when facing a problem.

Many times a piece is created whose final form is different than what was initially designed. As one of the project artists, Roxane Riva, said, "It's really good to see what you have instead of what you wanted." Oftentimes the end result is better than what was planned.

• **Some errors can be resolved by removing a section of enamel.**

The easiest way to remove enamel is to use a diamond burr on a flexible shaft; a diamond hand file also works in some instances. An alundum stone can also be used. Be certain to use under running water so heat does not build up, and to wash away residue. In all cases, after removing enamel, wash well and scrub surface with a glass brush or scrubbing pad. The piece can now be reenameled and finished.

• *Problem: There are speckles in the enamel colors.*
Possible Cause: The enamel is not clean.
Possible Solution: See Cleaning Enamels page 23. If color cannot be cleaned, just use for counter enamel.

Possible Cause: The enamel has deteriorated.
Possible Solution: Test-fire and see if results are similar. If so, only use for counter enamel.

Possible Cause: The enamel was not dry enough before placing the piece in the furnace, and the grains jumped to another location due to the moisture turning to steam.
Possible Solution: Allow the enamel to dry completely before firing.

Possible Cause: The water being used to clean the brush for wet-packing is dirty.
Possible Solution: Change the water and clean brush.

Possible Cause: Enamel contains foreign matter.
Possible Solution: Grind out and reenamel.

Possible Cause: There was grinding stone or metal residue left over from stoning the piece.
Possible Solution: Grind out and reenamel.

Possible Cause: If using a metal that oxidizes, the edges were not cleaned between firings and the firescale jumped onto the enamel.
Possible Solution: Clean edges of the piece and continue.

Possible Cause: Firescale from trivet sloughed off onto the enamel.
Possible Solution: Clean trivets.

Possible Cause: The fire bricks in the kiln have deteriorated and bits have gotten into the piece.
Possible Solution: Vacuum the bricks.

• *Problem: The opaque enamel has black spots.*
Possible Cause: The piece was fired at too high a temperature or for too long.
Possible Solutions: Refire at a lower temperature, or add a new layer of color and fire at a lower temperature. Grind out and reenamel. Enamel opaques, especially reds, last. If this is not possible, then underfire them until the end. If enameling on copper, reds, yellow-oranges, and oranges can react with the copper. Fire colored enamel over a layer of flux. Try using a finer mesh enamel.

• *Problem: The fired piece has bubbles or pits in it.*
Possible Cause: The enamel was not dried properly.

Possible Solution: Grind out the bubbles. See Wet-packing on page 32, taking special notice of wicking out water and drying completely.

Possible Cause: Improper metal was used. Maybe it was a porous casting or a rolled piece of metal clay and pickling salts were trapped in it.

Possible Solution: Start over with proper metal for enameling. Be certain to rinse well after pickling, and do not pickle castings if the metal is porous.

Possible Cause: The enamel layer was too thick and the gases could not escape during firing.
Possible Solution: Grind out the questionable areas and reenamel.

Possible Cause: Too much holding agent was used or it was not dry enough.
Possible Solution: Grind out the questionable areas and reenamel.

Possible Cause: The enamel has deteriorated.
Possible Solution: Grind out the questionable areas and reenamel.

Possible Cause: Piece was fired at too high a temperature.
Possible Solution: Grind out the questionable areas, reenamel, and refire at a lower temperature.

Possible Cause: The layer of enamel just fired was too thin. There are no pits, just incomplete coverage.
Possible Solution: Add more enamel and fire again.

Possible Cause: If using fines, incomplete coverage may result, which looks like pits, but is vacant spaces.
Possible Solution: Add more enamel to make a thicker layer. Refire. If the bubbles cannot be removed, try matting the surface, which may disguise them; or try covering with foil.

• **Problem: The piece has buckled.**
Possible Cause: There are uneven coats of enamels between the front and back, or piece isn't counter-enameled.
Possible Solution: Fire as usual. When the item is taken out of furnace, immediately place it on a clean heat-absorbent surface and press top of piece with a planche; remove quickly to avoid cracking. Piece should now be flat. Once it is completely cooled, clean with water and a glass brush and continue with the next layer of enamel or counter enamel. After flattening, piece must be fired at least one more time to relieve stress caused by heat-forming.

Possible Cause: The piece has been fired at too high a temperature.
Possible Solution: See flattening solution above.

Possible Cause: Piece has not been properly stilted. For example, if making a light switch plate, it is best to add support in the center, as supporting it only on the edges will cause it to warp from its own weight.
Possible Solution: See flattening solution above.

• **Problem: The enamel looks uneven.**
Possible Cause: The enamel was not fired long enough.
Possible Solution: Return to the furnace and fire again.

Possible Cause: The layer was not even.
Possible Solution: Add more enamel and fire.

• **Problem: After firing the enamel turned a different color.**
Possible Cause: If using directly on silver, reds, pinks, oranges, and yellows turn brown and opalescent white turns yellow. This is due to the silver imparting a yellow color to others.
Possible Solution: Basically, these colors need to be fired over a layer of flux for silver or fired over gold foil. Additionally, reds should be fired near the end of the piece at only 1350°F–1400°F. First, try to remove enamel with the unpleasant color. If this is impossible, hide the layer by covering with silver foil, then fire. This will change the look of the piece. Cover silver/silver foil with a layer of flux for silver, then fire. Reenamel and proceed.
Note: Some Japanese enamels have some colors especially formulated for silver.

Possible Cause: If transparent clear turned green on copper, then there was too thin a layer of enamel or it was fired at too hot a temperature or for too long.
Possible Solution: Next time, either apply a heavier layer or don't overfire; be aware of split-second timing.

Possible Cause: If using copper, you may be seeing a reddish oxide that is under the enamel. This will have a "tomato-ish" color.
Possible Solution: Refire at a higher temperature.

Possible Solution: Refire at a higher temperature or for a longer period of time. The oxide may be absorbed into the enamel and not show.

Possible Cause: Metal may have been dirty.
Possible Solution: Clean metal better next time.

Possible Cause: The piece may be overfired.
Possible Solution: Remove enamel and reenamel. Or fire-on a layer of silver or gold foil and reenamel on top of that. However, if the layer that turned color is too thin, add some extra enamel and fire before using foil. Some colors change over progressive firings. For example, transparent reds turn opaque. Fire these colors only on the last few firings.

• *Problem: The opaque enamel turned transparent.*
Possible Cause: The piece has been fired for too long and/or too high. This can happen particularly with gray, blue, turquoise, and green lead-bearing opaques. Sometimes there will be a very nice shading from opaque to transparent; but if it is to be saved, that must be the last firing or it will be lost on the next one.
Possible Solution: Fire again at a lower temperature.

• *Problem: The enamel just popped or chipped off or is cracking.*
Possible Cause: The metal was not clean.
Possible Solution: Using the piece that broke off, clean the metal in the exposed area, then reapply the broken chip with holding agent. Dry and fire.

Clean the metal, then build up the layers again to fill in the space.

Possible Cause: Counter enamel may not have been applied yet.
Possible Solution: Counter-enamel during the next firing.

Possible Cause: The counter enamel and enamel on the front are not the same thickness.
Possible Solution: Try stoning some enamel off and refire. If there are deep crevices in the counter enamel, fill them by mixing -200 mesh enamel with straight holding agent. Brush holding agent into crevices and then wet-pack the -200 mesh/holding agent combination. Using a sprayer, spray holding agent into the air above piece so a thin coating floats down onto the piece. Sift enamel over the entire side, dry completely, and fire.

Possible Cause: The last layer of enamel was not even or was too thick.
Possible Solution: Even out the layer, be certain to counter-enamel and refire.

Possible Cause: The metal is not an even thickness or has areas that are too narrow.
Possible Solution: Use the proper metal and don't cut areas that will create stress points.

Possible Cause: If using copper, look to see if the back of the chip that came off has firescale. If so then the piece was put into furnace at too low a temperature and/or not fired long enough.
Possible Solution: Clean, reapply enamel, and fire, being certain to put the piece in at the correct temperature.

Possible Cause: The enamel is not compatible with the metal's expansion.
Possible Solution: Use a different metal or a different enamel.

Possible Cause: If this occurs during the finishing process, then handling was too rough.
Possible Solution: Clean the piece completely with a glass brush. Then try to repair by either reenameling or using holding agent to help set the piece that cracked off.

Possible Cause: If using cloisonné wires, stress can be caused by certain designs of the wire like sharp points.
Possible Solution: Try refiring, but in general there is nothing to be done; it is a design issue.

• *Problem: The transparent enamel looks cloudy.*
Possible Cause: The enamel was not washed properly.
Possible Solution: See Cleaning Enamels page 23.

Possible Cause: The layer is too thick.
Possible Solution: Grind off and enamel a thinner layer, or cover with foil, fire, then enamel a thinner layer. Try firing at a high temperature (1500°F–1550°F).

Possible Cause: Some solder ghost was under the enamel.
Possible Solution: Clean better next time. Try covering with foil, fire, and reenamel.

Possible Cause: Pickle was left on a copper piece.
Possible Solution: Be certain to rinse piece well after pickling. Scrub with a glass brush before enameling.

Possible Cause: Enamel has deteriorated.
Possible Solution: Grind out or cover with foil and use different enamel.

Possible Cause: Used too much holding agent or the holding agent went bad.
Possible Solution: Grind out or cover with foil and use different enamel.

• *Problem: A color from a previous layer is showing through a top layer.*
Possible Cause: This is called pull-through or break-through. This can be an interesting effect. It is caused by a sufficiently high surface tension of some enamels. At the firing temperature the viscosity is reduced enough to permit the surface tension to take over and prevent the complete spreading of the top enamel, providing discontinuities for the first coat to flow into.
Possible Solution: Nothing can be done short of changing enamel. To minimize this effect, use -150/+325 mesh size particles as a second coat. Fire at a lower temperature for a longer time.

Possible Cause: Piece fired at too high a temperature.
Possible Solution: Enamel again and fire at a lower temperature.

Possible Cause: Top layer was too thin.
Possible Solution: Apply thicker layer of enamel.

Possible Cause: A lead-based enamel was used under a lead-free enamel.
Possible Solution: Only use lead-free under leaded, not vice versa.

• *Problem: Finished enamel piece was dropped and it cracked.*
Possible Cause: Enamel piece was dropped.
Possible Solution: Clean thoroughly, reenamel areas that flaked off and refire. If piece is old and very dirty, clean with ammonia and rinse very well (be certain to work in a well-ventilated area).

• *Problem: The cloisonné wires melted.*
Possible Cause: This will only happen when using a copper base with silver cloisonné wires and firing at too high a temperature. Silver and copper form a new alloy, which melts at a lower temperature than either of the two metals.
Possible Solution: Keep firing temperature down. Drill out melted wire and start over.

Make certain flux undercoat fired clear before adding wires.

Make certain wires are not directly touching copper backing; keep them suspended in the enamel layer.

• *Problem: The enamel has disappeared from the edges.*
Possible Cause: The layer was too thick.
Possible Solution: Brush some holding agent onto the bare areas and reapply enamel in a thinner coat; fire.

Possible Cause: The metal was not clean or pickle was not removed properly.
Possible Solution: Clean properly, apply some holding agent onto the bare areas and reapply enamel. Try scrubbing with a glass brush first.

Possible Cause: Enamel pulls away from the greater mass. If particle size of the enamel is too small, too heavy of an application will result with the enamel pulling away from the edge.
Possible Solution: Apply thinner coats of enamel or grade-sift the enamel with a 325 or 200 mesh sifter.

• *Problem: There is white or iridescent film on enamel.*
Possible Cause: Pickle was not removed properly.
Possible Solution: Scrub with a glass brush.

Drill out the area and reenamel.

This may disappear with firing.

On sterling, fine silver can be brought to the top by annealing and pickling multiple times until the metal does not turn black when heated. Bringing the fine silver to the top will mean that it won't be necessary to pickle after the last firing.

Try to remove with glass polishing compound on a motorized buffing wheel.

Possible Cause: This happens to acid-sensitive enamels.
Possible Solution: Use fine silver instead of copper so pickling isn't necessary, or use a lead-free, acid-resistant enamel.

Possible Cause: The enamel was not clean enough.
Possible Solution: Wash enamels better.

Note: Never use hot pickle on enamels. Lead-bearing enamels are less acid-resistant than lead-free. Therefore, lead-bearing enamels should be watched more closely when pickling.

Dedication

I dedicate this book to my mother, Renée Cohen, who always encouraged my artwork; my daughter, Judith Lanza, who hopefully is inspired by my encouragement of her artwork; my sister, Susan J. Rochlin, and my husband, Stephen Lanza, for all their support; and to Marilyn Druin, a wonderful enamelist who recently passed away, and who inspired me when I started enameling again after a twelve-year hiatus.

Appendix

Bibliography

Assorted Pearls & Gems, Enamel Guild South's Book of Tips on Enameling, © 1991 by Enamel Guild South, Inc. To order, contact Donna Buchwald, 8100 SW 92 Avenue, Miami, FL 33173, (305) 595-5767

Cloisonné Enameling and Jewelry Making, by Felicia Liban & Louise Mitchell, Dover Publications, Inc., New York, © 1980, ISBN: 0-486-25971-4

Complete Metalsmith, The, An Illustrated Handbook, by Tim McCreight, Davis Publications, Inc., Worcester, MA, © 1991, ISBN: 0-87192-240-1

Dictionary of Enameling, History and Techniques, by Erika Speel, Ashgate Publishing Limited, England, © 1998, ISBN: 1-85928-272-5

Enamel, Enameling, Enamelists by Glenice E. Mathews, Chilton Press, ISBN: 080197285X, 1984

Enameling - Principles & Practice, by Kenneth F. Bates, The World Publishing Company, Cleveland and New York, © 1951

Experimental Techniques in Enameling, by Fred Ball, VA, Nostrad Reinhold, NY 144 pp 1972

First Steps in Enameling by Jinks McGrath, The Wellfleet Press, A Division of Book Sales, Inc., Edison, New Jersey, © 1994, ISBN: 0-7858-0033-6

Glass on Metal, monthly magazine published by The Enamelist Society, PO Box 310, Newport, KY 41072, phone: (859) 291-3800

Hydraulic Die Forming for Jewelers and Metalsmiths, by Susan Kingsley, © 1991, ISBN: 0963583204

Metal Techniques for Craftsmen, by Oppi Utrecht, Doubleday & Co., Inc, ISBN: 0-385-03027-4, © 1968

Step by Step Enameling, A Complete Introduction to the Craft of Enameling, by William Harper, © 1973, Golden Press, New York

The Thames and Hudson Manual of Etching and Engraving, by Walter Chamberlain, © 1972, Thames and Hudson LTD, London

Thompson Enamel Workbook, Edited by Tom Ellis. To order, contact Thompson Enamel, Inc., PO Box 310 Newport, KY 41072, (859) 291-3800

Tudor, Jean, articles: "Raku Enameling," *Glass on Metal*, vol. 20, no. 3, June 2001, 62–65. And "Omission in Glass on Metal," *Glass on Metal*, vol. 20, no. 4, August 2001, 85.

Professional Organizations

The Enamelist Society, PO Box 310, Newport, KY 41072; (859) 291-3800

Enameling information

Other pertinent information on enameling can be found on Karen L. Cohen's web site at:

http://www.kcEnamels.com

Acknowledgments

When I was first asked to write this book, I knew it would take a lot of work. Well the work was worth it—I had a lot of fun, expanded my enameling knowledge and made quite a few friends. It was a great project to work on, and I couldn't have done it without the help and support of the project artists. I'd like to thank all of them for their contributions to the technical information, help with chapter reviews and their hard work on their projects. In addition, I would like to thank the following people whose extra help made this book what it is.

When I got this assignment, my first call was to June Jasen, a friend from my local guild, who helped me get started with finding other artists and then acted as a technical editor. Katharine Wood was wonderful for my morale and also helped with finding other artists. Also, her technical contributions included a write-up on mesh sizes. Charles Lewton-Brain was a great resource of information and offered resources to find information. He was always there when I needed something and I appreciate all he did. Jean Tudor was very helpful when I had questions, in reviewing the Troubleshooting section, and producing sample tiles and illustrative photos. I spent a lot of time on the phone with Coral Shaffer and would like to thank her for her support throughout the writing of this book. Tom Ellis helped me in many ways, including his write-up and references on various topics, extra photos, and help as technical editor. Diane Alymeda was very helpful with reviewing the section on wire bending and providing me with extra illustrative photos. Ora Kuller has been really terrific with her support and information and photos on particle sizes, including the chart of when to use what size. Judy Stone was especially helpful with her support and review of particle sizes. Sally Wright provided extra information on sifting, which was helpful. Dee Fontans contributed her write-up on how to fuse a bezel. Ralph Gabriner has my thanks for doing such a great job in photographing most of the final project pieces, and some of the step-by-steps. And lastly, but certainly not least, I would like to thank my editor, Ray Cornia, who was just great to work with. I love his book design and appreciate the support he gave me.

Thank you all.

Conversion Tables

MM-Millimeters CM-Centimeters
Inches to Millimeters and Centimeters

Inches	MM	CM	Inches
⅛	3	0.3	9
¼	6	0.6	10
½	13	1.3	12
⅝	16	1.6	13
¾	19	1.9	14
⅞	22	2.2	15
1	25	2.5	16
1¼	32	3.2	17
1½	38	3.8	18
1¾	44	4.4	19
2	51	5.1	20
2½	64	6.4	21
3	76	7.6	22
3½	89	8.9	23
4	102	10.2	24
4½	114	11.4	25
5	127	12.7	26
6	152	15.2	27

Solid Measures

Ounces to Grams		Pounds to Kilograms	
oz.	g.	lbs.	kg.
1	28.35	1	.4536
2	56.7	2	.907
3	85.05	3	1.361
4	113.4	4	1.814
5	141.75	5	2.268
6	170.1	6	2.722
7	198.45	7	3.175
8	226.8	8	3.629
9	255.15	9	4.082
10	283.5	10	4.536
11	311.85	11	4.99
12	340.2	12	5.443
13	368.55	13	5.897
14	396.9	14	6.350
15	425.25	15	6.804

Liquid Measures

Ounces to Millilitres	
oz.	ml.
¼	7
½	15
1	28
2	56
3	85
4	110
5	140
6	170
7	196
8	225

tsp. = teaspoon
Tbs. = tablespoon
3 tsp. = 1 Tbs.
16 Tbs. = 1 cup
1 cup = 8 oz.

Thickness Measures

Inches to Gauge	
inch	B&S ga
.003	40
.005	36
.008	32
.010	30
.012	28
.016	26
.020	24
.025	22
.032	20
.040	18
.051	16
.064	14
.081	12

Photo Credits

Almeyda, Diane Echnoz: 19 lr, 34 ll

Anthony, Jerry: 93 all

Balazs, Harold: 6 lr, 11 ctr

Bankenroth, Adam: 145 lr

Barrett, Bob: 1, 10 ul

Bauser, Jennifer: 20 ll, 53 all, 54 all, 55 ur, ctr, 68 ll, ctr, 69 all, 70 ul

Berg, Geraldine M.: 71 ul

Brady, Jeff: 6 ll

Brannon, Rebecca: 12 2nd from lower

Browne, Kathleen: 105 l ctr

Brunck, Richard: 18, 151 ll

Bryan, Allen: 133 ul

Butt, Harlan W.: 4

Cameron, Kate: 7 ll

Clark, Ann: 37 ctr

Cole, Maureen: 105 ll

Conrad, Ernst: 124 lr, 125 all

Corwin, Isabella: 71 lr

Continued . . .

Darty, Linda: 2 lr, 45 lr, 151 lr

De Geus, Jack: 25 lr

Fillion, Norman: 45 ul

Fontans, Dee: 60

Fuhrman, Leni: 11 ur

Gabriner, Ralph: 2 ul, 7 ul, ur, 8 ll, 9 ur, lr, 34 ur, 37 ll, lr, 42 ul, 50 lr, 51 all, 52, 55 lower, 68 ur, 70 ll, 102 all, 103 ur, ctr, 104 all, 105 lr, 107 upper, 108, 111, 112 lower, 113 all, 114 top, 115 lr, 116 lower, 117 ul, lower, 118 ul, 120 lr, 121 ul, 123 ll, 126 ul, 127 lr, 145 ll, 150 lr

j.e.jasen: 12 ul

Killmaster, John: 6 ur

Komrad, Audrey B.: 7 r, 151 ul

Kuller, Ora: 21 lr, 22, 25 ur,

Lanza, Stephen: 13 all, 14 all, 15 all, 16 all, 17 ul, ll, 19 ur, 20 u series, 21 u, 23, 24 all, 26, 27 all, 28 ur, ctr, 29, 30 all, 31 all, 32 ll, ctr, 33 all, 35, 36 all, 37 ur, 38 all, 39 all

Laskin, Rebekah: 12 lr, 129 ur

Lemke, Bill: 40 ll

Lewton-Brain, Charles: 128 lower, 129 ul, l ctr, 130 all, 131 all, 132 all

Loffler, Dan: 3, 100 all, 101 all

Lozier, Deborah: 11 ll

MacKarell, Joan: 65 ll, lr

Marraccini, Pam: 87 ul

McFadyen, Barbara: 45 ll

Miller, Christina T.: 105 upper ctr

Mills, Tom: 146 ll, 147 all

Nickel, Joel: 107 ll, lr

Post, George: 8 ul, ml, 10 upper ctr, 65 ur, 71 ur, 88, 89, 90 all, 91, 145 ul

Peterson, Diane: 140 all, 141 all, 142 all, 143 all, 144 all

Preston, Gregory: 133 ll

Rae, MerryLee: 9 lr

Ratz, W. Doris: 105 ur

Riva, Raymond: 135 ul, 137 all, 138 ul, ctr, ll,

Robb, Jim: 2 top

Rohman, Gary: 134 lower, 135 ur, lr, 136 ul, 139 lr, 139 all

Rooke, Fay: 10 ll, ur

Rooke-Harris, Diane: 12 2nd from upper, 104 ul

Sakwa, Hap: 2 ll, 71 ll, 86 lr, 87 ur, lower, 151 ur

Schreiber, Roger: 66 lr, 67 ul, ll

Schwed, Antonia: 2nd from lower

Scovil, Jeff: 65 ul

Shepps, Anton: 17 ctr, 92 lower, 94 all, 95 all

Silverman, Paul: 46 lr, 47 upper

Slovil, Jeff: 9 2nd from upper

Sommer, Tom: 65 l ctr

Tanzer, Jo Ann: 11 ul

Tudor, Jean: 11 lr, 32 ul

Turell, Elizabeth: 133 ur

Wetmore, Debbie: 145 ur

Wood, Katharine: 116 ul

Wright, Gerald: 117 ur

Wright, Robert: 118 ctr, ll, 119 all, 120 ul, ctr, ll, 121 ll, ctr, lr, ur, 122 all, 123 ur, 2nd from upper, 2nd from lower, lr

Wright, Sally: 28 lr, 46 ul

Zilker, Jack: 133 lr

Index